The Behavioural
Biology of Chickens

The Behavioural Biology of Chickens

Editor

Surendra Mohite

The Behavioural Biology of Chickens

Edited by **Surendra Mohite**

Printed in 2017

ISBN: 978-1-68117-100-5

Library of Congress Control Number: 2015935500

© 2016 by

SCITUS Academics LLC,
616, Corporate Way, Suite 2, 4766,
Valley Cottage, NY 10989

www.scitusacademics.com

Contents

vi

Preface

Archaeological evidence suggests that the bird commonly known as the chicken (Gallus domesticus) is a domesticated version of the Indian and Southeast Asian Red Jungle Fowl (Gallus gallus) which is still found in the wild today. It is thought that the bird was first tamed in China around 6000 BC, with the birds moving into India by 2000 B.C. The chicken then spread from China to Russia and from there into Europe between 750 B.C. – 42 A.D. Some scholars believe that the bird may have been domesticated first for its use in cockfighting, and only later used as a food source. The White Leghorn breed or crosses of this breed is the large white bird most commonly used in agriculture and for research, but there are over 400 different breeds of chickens. Chickens have a rigid social structure called the "pecking order" by which every bird establishes who is dominant and who is submissive in relationship to every other bird. Dominant birds peck at submissive birds, pluck their feathers, and may chase them away or steal their food. Submissive birds will not peck back and will usually run from the dominant birds. Anytime a bird is added or subtracted from the flock, even if it is only a well-known bird that has been temporarily removed and then returned to the group, the entire flock will fight briefly to re-establish the pecking order. Flocks of greater than 15 birds can lead to excessive fighting and less productivity. Males should not be kept together as they will often fight each other and may even sexually abuse or kill the weaker birds.

Editor

Chapter 1

The Possible Role of the Uropygial Gland on Mate Choice in Domestic Chicken

Atsushi Hirao

Division of Anatomy & Embryology, Department of Anatomy, School of Medicine, Jichi Medical University, Shimotsuke, Tochigi 329-0498, Japan

ABSTRACT

In avian mating systems, male domestic fowls are polygamous and mate with a number of selected members of the opposite sex. The factors that influence mating preference are considered to be visual cues. However, several studies have indicated that chemosensory cues also affect socio-sexual behavior, including mate choice and individual recognition. The female uropygial gland appears to provide odor for mate choice, as uropygial gland secretions are specific to individual body odor. Chicken olfactory bulbs possess efferent projections to the nucleus taeniae that are involved in copulatory behavior. From various reports, it appears that the uropygial gland has the potential to act as the source of social odor cues that dictate mate choice. In this review, evidence for the possible role of the uropygial gland on mate choice in domestic chickens is presented. However, it remains unclear whether a relationship exists between the uropygial gland and major histocompatibility complex-dependent mate choice.

INTRODUCTION

Nearly all mammals emit chemical substances into their surroundings and these substances have important effects on mating behavior. For example, male house mice (Mus musculus) scent mark with urine to attract females for mating. Additionally, female mice are able to distinguish between the odors of parasitized and unparasitized males and are attracted to the odor of the latter [1–3]. It appears the odors that these mating preferences evoke can be attributed to the major histocompatibility complex (MHC) [4].

In contrast to mammals, and as avian species have often been classified as anosmic or microsmatic [5–9], olfactory information is generally not considered to be involved in the mating behavior of birds. However, several investigators have suggested that chemical cues, such as individual recognition and mate choice, affect avian social behavior [5–8, 10, 11]. In addition, the Blue Tit (Cyanistes caeruleus L.) can detect chemical secretions of predators and exhibit antipredatory behavior to reduce the risk of predation [12]. More recently, it has been reported that the female chicken (Gallus gallus domesticus) uropygial gland is related to male mate choice [13]. Mate choice is defined as any pattern of behavior, shown by members of one sex, which leads to them being more likely to mate with certain members of the opposite sex than with others [14].

The three aims of this review are to present the factors that evoke mate choice in domestic chickens, examine the possible role of the chicken uropygial gland as a source of social odor cues, and discuss whether uropygial gland secretions affect MHC-dependent mate choice.

WHAT SIGNALS ELICIT MATE CHOICE IN DOMESTIC CHICKENS?

Mating behavior in domestic chickens has been described in detail by previous investigators [15]. Prior to mating, a series of courtship displays take place before mating based on a stimulus-response sequence initiated by males (Figure 1). Furthermore, several researchers have provided supporting evidence that domestic fowls exhibit non-random mating [16–18].

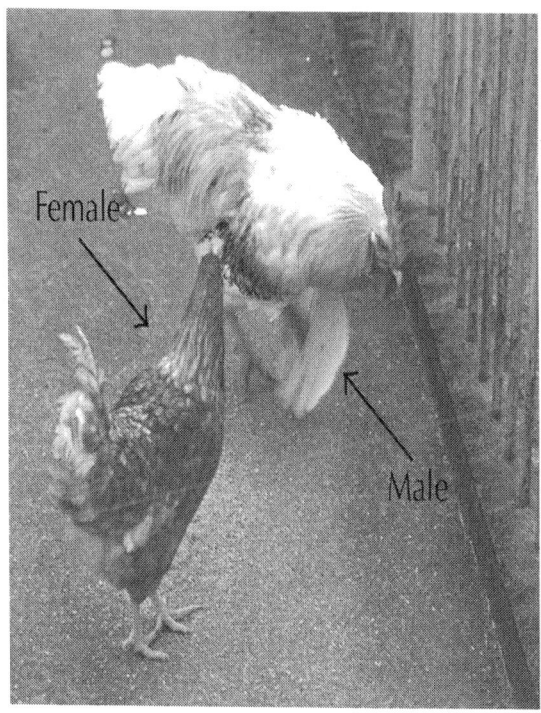

(a)

(b)

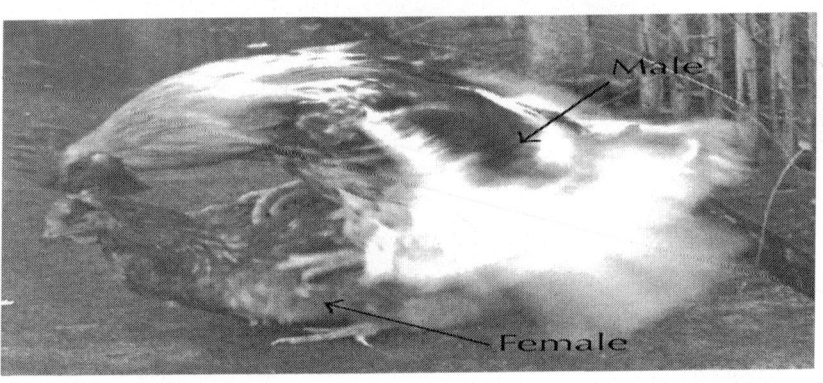

(c)

Figure 1: Photographs of sexual behavior exhibited by domestic chickens. (a) Courtship waltzing, (b) mounting, and (c) copulation.

In domestic fowls, vision appears to play a central role in mating behavior [19, 20]. As the size of sexual ornaments, such as combs, wattles, and spurs are under the control of testosterone [21], these ornaments are regarded as signals affecting mate choice. Zuk et al.

[22] and Johnsen and Zuk [23] suggested that longer, redder combs in male red jungle fowls (Gallus gallus) were preferred by females of the species. Graves et al. [24] also reported that male chickens having lager combs were selected more often by female birds. A recent study on male wattles reported that male wattle size significantly reduces orienting latency in tidbitting display [25]. From these reports, the comb and wattle size of male chickens and red jungle fowls appears to act as a dominant signal influencing mate selection by female birds.

On the other hand, there is little evidence that chemical signals are involved in mate choice in domestic chickens. Recently, however, it has been further suggested that the female uropygial gland provides an olfactory cue mediating mate choice. For instance, male domestic chickens mate significantly more with female birds possessing uropygial glands than with uropygial-glandectomized females [13]. Additionally, this mate preference disappeared in males subjected to olfactory sensory deprivation [13]. Thus, to investigate secretions from the uropygial gland as the source of odor cues merits further study.

IS THE CHICKEN UROPYGIAL GLAND A SOURCE OF SOCIAL ODOR?

Avian species with scent glands that emit strong odors are rarely observed. Thus, it is generally considered that birds do not use chemical information in mating behavior. However, birds possess the relatively large uropygial gland at the base of their tail feathers (Figure 2) [9, 26–28] which produces a large amount of volatile and nonvolatile compounds in the form of a waxy fluid that is spread on feathers as a part of plumage maintenance [9, 26–28]. Furthermore, volatile compounds in uropygial gland secretions exhibit seasonal changes [29–32]. A few recent studies have suggested that gland secretions include socio-ecological information, which allows distinction of species, gender, and even individuals [33, 34]. Moreover, several reports have shown that volatile compounds in uropygial gland secretions are responsible for odors with specific functions [32, 35, 36]. For example, the gland secretions of some birds contain volatile compounds that contribute to an unpleasant odor emitted to aid in the escape from predators [37]. Taken together, these reports suggest that volatile compounds in uropygial gland secretions act as chemical cues, and may reflect the social status of birds.

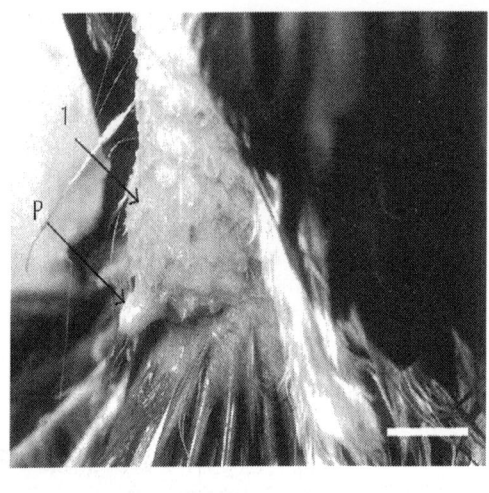

(a)

(b)

Figure 2: Photographs of the domestic chicken uropygial gland. (a) Lateral and (b) dorsal view of an adult uropygial gland. L: lobe, p: papilla. Scale bars indicate 1 cm.

In the case of domestic chickens, uropygial gland secretions rarely contain waxes [28], which are fundamental to waterproofing and maintaining the flexibility of feathers [28]. This finding suggests that the secretions possess another function besides waterproofing. Indeed, the red jungle fowl emits an individual body odor that is produced by aliphatic carboxylic acids [38]. Moreover, trained mice are able to discriminate between these odors at the level of the individual [38]. Based on this evidence, chicken uropygial gland secretions have the potential to function as social odor cues.

UTILIZATION OF OLFACTORY CUES FOR MATING SYSTEMS IN DOMESTIC CHICKEN

In contrast to previous works on avian olfactory function, electrophysiological studies have provided evidence that domestic fowls are indeed capable of perceiving odor cues. For example, chicken olfactory bulbs respond to odor stimuli [39, 40]. In addition, an in situ hybridization study revealed that a number of olfactory receptor genes have been characterized in the olfactory epithelium [41]. Recently, a second class within the odorant receptor family, termed trace amine-associated receptors (TAARs), was identified. Certain mouse TAARs are able to perceive volatile amines present in urine, and one TAAR was found that recognizes a pheromone compound [42, 43]. From these results, it is suggested that one function of TAARs involves the detection of social cues [42, 43]. Moreover, database searches have revealed that domestic chickens possess three functional TAAR genes [44], and a protein sequence of chicken TAARs has also been determined [45]. More recently, Gomez and Celii [46] have established a culture method of olfactory sensory neurons, which is a powerful tool for in vitro studies aimed at understanding olfactory perception in domestic chickens.

Olfactory bulbs of domestic chickens are innervated by efferent fibers [47] and possess similar projection sites to that of other birds [48–50]. Moreover, chicken olfactory bulbs project to the nucleus taeniae [47]. In Japanese quail (Coturnix japonica), a lesion of this nucleus causes a significant reduction in the frequency of copulation [51]. Based on these findings, domestic chickens appear to possess functional olfactory systems that influence mating behavior.

Behavioral investigations have also demonstrated that domestic chickens react to various olfactory stimuli [52]. It seems that chemical information plays an important role for their life. However, the direct evidence that domestic chicken might use chemosensory cues to assess mating behavior is rarely reported. In other birds, such as mallard ducks (Anas platyrhynchos), bilateral olfactory nerve sectioning significantly reduced the number of social and mating behavior [53]. In Japanese quail, bilateral nostril sealing decreased the number of mating behavior [6, 54]. To understand the role of olfaction in mating behavior, it is at least necessary to perform similar experiments in domestic chickens.

IS MHC-DEPENDENT MATE CHOICE IN CHICKENS MEDIATED BY THE UROPYGIAL GLAND?

In mice, MHC-based mate selection is proposed to involve the detection of male odors by females that leads to mating with males carrying dissimilar MHC genes, and results in progeny with disease-resistance genes [55–60]. In avian species, although a few investigators have suggested that mate choice might be affected by olfaction [32, 61, and 62], there is little evidence for the direct relation between MHC-dependent mate choice and the uropygial gland.

However, recent studies suggest the possibility that MHC genes are related to mate choice in birds. According to research of outbred populations, house sparrows (Passer domesticus) appear to exhibit MHC-based mate choice [63]. Moreover, female house sparrows seem to utilize olfactory cues for MHC-dependent mating preference [64]. Male red jungle fowls show several cryptic preferences by allocating additional sperm to MHC-dissimilar females [65]. Additionally, it has been shown in several bird species that uropygial gland size changes with the load of feather mites, bacteria, and chewing lice [66–68], while removal of the uropygial gland leads to increased levels of fungi and feather-degrading bacteria on feathers, and higher levels of feather degradation [28]. These findings are supported by a study that demonstrated that chicken uropygial gland secretions reduce the levels of these microorganisms on feathers [69]. Taken together, these reports suggest that chemical defenses provided by the uropygial gland may reflect the status of disease-resistance. It is assumed that uropygial gland secretions contain MHC proteins. Unfortunately, this possibility has not been explored. The issue should be examined to understand MHC-dependent mate choice in domestic chickens.

CONCLUSIONS AND FUTURE WORK

It is known that male domestic chickens prefer to mate with certain members of the opposite sex, with previous works suggesting that visual cues play a central role in mating behavior. Undoubtedly, domestic chickens depend predominantly on visual information to function, while olfaction appears to play a role in their life. Chemical cues from the uropygial gland may compensate for information that vision is not able to detect.

Finally, future investigations on the uropygial gland and mate choice in domestic chickens should consider two important issues. Firstly, although MHC genes heavily affect mate choice [4] in mammals through olfaction, it remains unclear whether uropygial gland secretions contain MHC proteins. Resolving this issue is necessary to understand mate choice in domestic chickens. Secondly, the localization of olfactory receptors which are able to perceive social odor cues has not been examined. For instance, mouse V2 receptors are able to perceive odor substances in urine and therefore play an important role in MHC-dependent mate choice [4]. To determine the localization of such olfactory receptors in domestic chickens, it is first necessary to elucidate the mechanisms of perceiving social odor.

REFERENCES

1.   M. Kavaliers and D. D. Colwell, "Aversive responses of female mice to the odors of parasitized males: neuromodulatory mechanisms and implications for mate choice," Ethology, vol. 95, no. 3, pp. 202–212, 1992.

2.   M. Kavaliers and D. D. Colwell, "Discrimination by female mice between the odours of parasitized and non-parasitized males," Proceedings of the Royal Society of B, vol. 261, no. 1360, pp. 31–35, 1995.

3.   M. Kavaliers and D. D. Colwell, "Odours of parasitized males induce aversive responses in female mice,"Animal Behaviour, vol. 50, no. 5, pp. 1161–1169, 1995.

4.   P. A. Brennan and K. M. Kendrick, "Mammalian social odours: attraction and individual recognition,"Philosophical transactions of the Royal Society of London. Series B, Biological sciences, vol. 361, no. 1476, pp. 2061–2078, 2006.

5.   S. P. Caro and J. Balthazart, "Pheromones in birds: myth or reality?" Journal of Comparative Physiology A, vol. 196, no. 10, pp. 751–766, 2010

6.   J. Balthazart and M. Taziaux, "The underestimated role of olfaction in avian reproduction?" Behavioural Brain Research, vol. 200, no. 2, pp. 248–259, 2009.

7.   J. C. Hagelin and I. L. Jones, "Bird odors and other chemical substances: a defense mechanism or overlooked mode of intraspecific communication?" Auk, vol. 124, no. 3, pp. 741–761, 2007

8.   J. C. Hagelin, "Odors and chemical signaling," in Reproductive Behavior and Phylogeny of Birds, B. G. M. Jamieson, Ed., pp. 76–119, Science Publishers, New Hampshire, NY, USA, 2007.

9.   T. J. Roper, "Olfaction in birds," in Advances in the Study of Behavior, P. J. B. Slater, J. S. Rosenblat, C. T. Snowden, and T. J. Roper, Eds., pp. 247–332, Academic Press, Boston, NY, USA, 1999.

10.   F. Bonadonna, "Olfaction in petrels from homing to self-odor avoidance," Annals of the New York Academy of Sciences, vol. 1170, pp. 428–433, 2009

11.   T. W. O'Dwyer and G. A. Nevitt, "Individual odor recognition in procellariiform chicks: potential role for the major histocompatibility complex," Annals of the New York Academy of Sciences, vol. 1170, pp. 442–446, 2009

12.   L. Amo, I. Galvan, G. Tomas, and J. J. Sanz, "Predator odour recognition and avoidance in a songbird,"Functional Ecology, vol. 22, pp. 289–293, 2008.

13.   A. Hirao, M. Aoyama, and S. Sugita, "The role of uropygial gland on sexual behavior in domestic chicken Gallus gallus

domesticus," Behavioural Processes, vol. 80, no. 2, pp. 115–120, 2009

14. T. R. Halliday, "The study of mate choice," in Mate Choice, P. Bateson, Ed., pp. 3–32, Cambridge University Press, Cambridge, UK, 1983.

15. A. M. Guhl and G. Fischer, "The behaviour of chickens," in The Behaviour of Domestic Animals, E. S. E. Hafez, Ed., pp. 515–553, Baillère Tindall, London, UK, 1975.

16. C. W. Upp, "Preferential mating in fowls," Poultry Science, vol. 7, pp. 225–232, 1928.

17. A. M. Guhl, "Measurable differences in mating behaviour of cocks," Poultry Science, vol. 30, pp. 687–693, 1951.

18. A. Lill, "Some observations on social organisation and non-random mating in captive Burmese red jungle fowl," Behaviour, vol. 26, pp. 228–242, 1966.

19. E. K. M. Jones, N. B. Prescott, P. Cook, R. P. White, and C. M. Wathes, "Ultraviolet light and mating behaviour in domestic broiler breeders," British Poultry Science, vol. 42, no. 1, pp. 23–32, 2001

20. E. K. M. Jones and N. B. Prescott, "Visual cues used in the choice of mate by fowl and their potential importance for the breeder industry," World's Poultry Science Journal, vol. 56, no. 2, pp. 128–138, 2000.

21. A. B. Gilbert, "The endocrine ovary in reproduction," in Physiology and Biochemistry of the Domestic Fowl, D. J. Bell and B. M. Freeman, Eds., pp. 1449–1468, Academic Press, London, UK, 1971.

22. M. Zuk, T. S. Johnsen, and T. MacLarty, "Endocrine-immune interactions, ornaments and mate choice in red jungle fowl," Proceedings of the Royal Society of B, vol. 260, no. 1358, pp. 205–210, 1995

23. T. S. Johnsen and M. Zuk, "Repeatability of mate choice in female red jungle fowl," Behavioral Ecology, vol. 7, no. 3, pp. 243–246, 1996.

24. H. B. Graves, C. P. Hable, and T. H. Jenkins, "Sexual selection in gallus: effects of morphology and dominance on female spatial

behavior," Behavioural Processes, vol. 11, no. 2, pp. 189–197, 1985.

25.   C. L. Smith, D. A. Van Dyk, P. W. Taylor, and C. S. Evans, "On the function of an enigmatic ornament: wattles increase the conspicuousness of visual displays in male fowl," Animal Behaviour, vol. 78, no. 6, pp. 1433–1440, 2009.

26.   A. Salibian and D. Montalti, "Physiological and biochemical aspects of the avian uropygial gland,"Brazilian Journal of Biology, vol. 69, no. 2, pp. 437–446, 2009.

27.   J. Jacob, "Uropygial gland secretion and feather wax," in Chemical Zoology, A. H. Brush, Ed., pp. 165–211, Academic Press, London, UK, 1978.

28.   J. Jacob and V. Ziswiler, "The uropygial gland," in Avian Biology, D. S. Frander, J. R. King, and K. C. Parks, Eds., pp. 199–324, Academic Press, New York, NY, USA, 1982.

29.   J. Jacob, J. Balthazart, and E. Schoffeniels, "Sex differences in the chemical composition of uropygial gland waxes in domestic ducks," Biochemical Systematics and Ecology, vol. 7, no. 2, pp. 149–153, 1978.

30.   T. Piersma, M. Dekker, and J. S. Sinninghe Damsté, "An avian equivalent of make-up?" Ecology Letters, vol. 2, no. 4, pp. 201–203, 1999

31.   J. Reneerkens, T. Piersma, and J. S. Sinninghe Damsté, "Sandpipers (Scolopacidae) switch from monoester to diester preen waxes during courtship and incubation, but why?" Proceedings of the Royal Society of B, vol. 269, no. 1505, pp. 2135–2139, 2002.

32.   H. A. Soini, S. E. Schrock, K. E. Bruce, D. Wiesler, E. D. Ketterson, and M. V. Novotny, "Seasonal variation in volatile compound profiles of preen gland secretions of the dark-eyed junco (Junco hyemalis)," Journal of Chemical Ecology, vol. 33, no. 1, pp. 183–198, 2007. View at Publisher · View at Google Scholar · View at PubMed

33.   J. Mardon, S. M. Saunders, M. J. Anderson, C. Couchoux, and F. Bonadonna, "Species, gender, and identity: cracking petrels› sociochemical code," Chemical Senses, vol. 35, no. 4, pp. 309–321, 2010

34. D. J. Whittaker, H. A. Soini, J. W. Atwell, C. Hollars, M. V. Novotny, and E. D. Ketterson, "Songbird chemosignals: volatile compounds in preen gland secretions vary among individuals, sexes, and populations," Behavioral Ecology, vol. 21, no. 3, pp. 608–614, 2010

35. M. Haribal, A. Dhondt, and E. Rodriguez, "Diversity in chemical compositions of preen gland secretions of tropical birds," Biochemical Systematics and Ecology, vol. 37, no. 2, pp. 80–90, 2009.

36. M. Haribal, A. A. Dhondt, D. Rosane, and E. Rodriguez, "Chemistry of preen gland secretions of passerines: different pathways to same goal? Why?" Chemoecology, vol. 15, no. 4, pp. 251–260, 2005.

37. B. V. Burger, B. Reiter, O. Borzyk, and M. A. Du Plessis, "Avian exocrine secretions. I. Chemical characterization of the volatile fraction of the uropygial secretion of the green woodhoopoe, Phoeniculus purpureus," Journal of Chemical Ecology, vol. 30, no. 8, pp. 1603–1611, 2004.

38. A. C. Karlsson, P. Jensen, M. Elgland et al., "Red junglefowl have individual body odors," Journal of Experimental Biology, vol. 213, no. 10, pp. 1619–1624, 2010. D. E. F. McKeegan and N. Lippens, "Adaptation responses of single avian olfactory bulb neurones,"Neuroscience Letters, vol. 344, no. 2, pp. 83–86, 2003.

39. T. Oosawa, Y. Hirano, and K. Tonosaki, "Electroencephalographic study of odor responses in the domestic fowl," Physiology and Behavior, vol. 71, no. 1-2, pp. 203–205, 2000. M. Leibovici, F. Lapointe, P. Aletta, and C. Ayer-Le Lièvre, "Avian olfactory receptors: differentiation of olfactory neurons under normal and experimental conditions," Developmental Biology, vol. 175, no. 1, pp. 118–131, 1996.

40. S. D. Liberles and L. B. Buck, "A second class of chemosensory receptors in the olfactory epithelium,"Nature, vol. 442, no. 7103, pp. 645–650, 2006.

41. S. D. Liberles, "Trace amine-associated receptors are olfactory receptors in vertebrates," Annals of the New York Academy of Sciences, vol. 1170, pp. 168–172, 2009.

42. Y. Hashiguchi and M. Nishida, "Evolution of trace amine-associated receptor (TAAR) gene family in vertebrates: lineage-specific expansions and degradations of a second class of vertebrate chemosensory receptors expressed in the olfactory epithelium," Molecular Biology and Evolution, vol. 24, no. 9, pp. 2099–2107, 2007.

43. J. C. Mueller, S. Steiger, A. E. Fidler, and B. Kempenaers, "Biogenic trace amine-associated receptors (TAARs) are encoded in avian genomes: evidence and possible implications," Journal of Heredity, vol. 99, no. 2, pp. 174–176, 2008.

44. G. Gomez and A. Celii, "The peripheral olfactory system of the domestic chicken: physiology and development," Brain Research Bulletin, vol. 76, no. 3, pp. 208–216, 2008.

45. A. Hirao, S. Sugita, and K. Sugahara, "Efferent and afferent connections and efferent pathway of olfactory bulb in the chicken (Gallus domesticus)," Animal Science Journal, vol. 71, pp. J483–J490, 2000 (Japanese).

46. P. Ebinger, G. Rehkamper, and H. Schroder, "Forebrain specialization and the olfactory system in anseriform birds. An architectonic and tracing study," Cell and Tissue Research, vol. 268, no. 1, pp. 81–90, 1992.

47. A. Reiner and H. J. Karten, "Comparison of olfactory bulb projections in pigeons and turtles," Brain, Behavior and Evolution, vol. 27, no. 1, pp. 11–27, 1985.

48. G. K. Rieke and B. M. Wenzel, "Forebrain projections of the pigeon olfactory bulb," Journal of Morphology, vol. 158, no. 1, pp. 41–55, 1978.

49. R. R. Thompson, J. L. Goodson, M. G. Ruscio, and E. Adkins-Regan, "Role of the Archistriatal Nucleus taeniae in the Sexual Behavior of Male Japanese Quail (Coturnix japonica): a Comparison of Function with the Medial Nucleus of the Amygdala in Mammals," Brain, Behavior and Evolution, vol. 51, no. 4, pp. 215–229, 1998.

50. R. B. Jones and T. J. Roper, "Olfaction in the domestic fowl: a critical review," Physiology and Behavior, vol. 62, no. 5, pp. 1009–1018, 1997

51. J. Balthazart and E. Schoffeniels, "Pheromones are involved in the control of sexual behaviour in birds,"Naturwissenschaften, vol. 66, no. 1, pp. 55–56, 1979.

52. M. Taziaux, M. Keller, G. F. Ball, and J. Balthazart, "Site-specific effects of anosmia and cloacal gland anesthesia on Fos expression induced in male quail brain by sexual behavior," Behavioural Brain Research, vol. 194, no. 1, pp. 52–65, 2008.

53. K. Yamazaki, E. A. Boyse, V. Mike, et al., "Control of mating preferences in mice by genes in the major histocompatibility complex," Journal of Experimental Medicine, vol. 144, no. 5, pp. 1324–1335, 1976.

54. K. Yamazaki, G. K. Beauchamp, D. Kupniewski, J. Bard, L. Thomas, and E. A. Boyse, "Familial imprinting determines H-2 selective mating preferences," Science, vol. 240, no. 4857, pp. 1331–1332, 1988.

55. D. J. Penn, "The scent of genetic compatibility: sexual selection and the major histocompatibility complex," Ethology, vol. 108, no. 1, pp. 1–21, 2002.

56. D. Penn and W. Potts, "MHC-disassortative mating preferences reversed by cross-fostering," Proceedings of the Royal Society of B, vol. 265, no. 1403, pp. 1299–1306, 1998.

57. W. K. Potts, C. J. Manning, and E. K. Wakeland, "Mating patterns in seminatural populations of mice influenced by MHC genotype," Nature, vol. 352, no. 6336, pp. 619–621, 1991.

58. C. R. Freeman-Gallant, M. Meguerdichian, N. T. Wheelwright, and S. V. Sollecito, "Social pairing and female mating fidelity predicted by restriction fragment length polymorphism similarity at the major histocompatibility complex in a song bird," Molecular Ecology, vol. 12, no. 11, pp. 3077–3083, 2003.

59. F. Bonadonna and G. A. Nevitt, "Partner-specific odor recognition in an antarctic seabird," Science, vol. 306, no. 5697, p. 835, 2004.

60. B. Zelano and S. V. Edwards, "An Mhc component to kin recognition and mate choice in birds: predictions, progress, and prospects," American Naturalist, vol. 160, no. 6, pp. S225–S237, 2002.

61. C. Bonneaud, O. Chastel, P. Federici, H. Westerdahl, and G. Sorci, "Complex Mhc-based mate choice in a wild passerine," Proceedings of the Royal Society of B, vol. 273, no. 1590, pp. 1111–1116, 2006

62. M. Griggio, C. Biard, D. J. Penn, and H. Hoi, "Female house sparrows "count on" male genes: experimental evidence for MHC-dependent mate preference in birds," BMC Evolutionary Biology, vol. 11, no. 1, article 44, 2011.

63. M. A. F. Gillingham, D. S. Richardson, H. Løvlie, A. Moynihan, K. Worley, and T. Pizzari, "Cryptic preference for MHC-dissimilar females in male red junglefowl, Gallus gallus," Proceedings of the Royal Society of B, vol. 276, no. 1659, pp. 1083–1092, 2009I. Galván, E. Barba, R. Piculo et al., "Feather mites and birds: an interaction mediated by uropygial gland size?" Journal of Evolutionary Biology, vol. 21, no. 1, pp. 133–144, 2008

64. A. P. Møller, G. A. Czirjak, and P. Heeb, "Feather micro-organisms and uropygial antimicrobial defences in a colonial passerine bird," Functional Ecology, vol. 23, no. 6, pp. 1097–1102, 2009

65. G. Moreno-Rueda, "Uropygial gland size correlates with feather holes, body condition and wingbar size in the house sparrow Passer domesticus," Journal of Avian Biology, vol. 41, no. 3, pp. 229–236, 2010

66. A. Bandyopadhyay and S. P. Bhattacharyya, "Influence of fowl uropygial gland and its secretory lipid components on the growth of skin surface fungi of fowl," Indian Journal of Experimental Biology, vol. 37, no. 12, pp. 1218–1222, 1999.

Chapter 2

Heritable Genome-Wide Variation of Gene Expression and Promoter Methylation between Wild and Domesticated Chickens

Daniel Nätt[1], Carl-Johan Rubin[2], Dominic Wright[1,] Martin Johnsson[1], Johan Beltéky[1], Leif Andersson[2], and Per Jensen[1]

[1]IFM Biology, Division of Zoology, Avian Behavioural Genomics and Physiology Group, Linköping University, Sweden

[2]Department of Medical Biochemistry and Microbiology, Uppsala University, Sweden

ABSTRACT

Background

Variations in gene expression, mediated by epigenetic mechanisms, may cause broad phenotypic effects in animals. However, it has been debated to what extent expression variation and epigenetic modifications, such as patterns of DNA methylation, are transferred across generations, and therefore it is uncertain what role epigenetic variation may play in adaptation.

Results

In Red Junglefowl, ancestor of domestic chickens, gene expression and methylation profiles in thalamus/hypothalamus differed substantially from that of a domesticated egg laying breed. Expression as well as methylation differences were largely maintained in the offspring, demonstrating reliable inheritance of epigenetic variation. Some of the inherited methylation differences were tissue-specific, and the differential methylation at specific loci were little changed after eight generations of intercrossing between Red Junglefowl and domesticated laying hens. There was an over-representation of differentially expressed and methylated genes in selective sweep regions associated with chicken domestication.

Conclusions

Our results show that epigenetic variation is inherited in chickens, and we suggest that selection of favourable epigenomes, either by selection of genotypes affecting epigenetic states, or by selection of methylation states which are inherited independently of sequence differences, may have been an important aspect of chicken domestication.

BACKGROUND

Chickens were domesticated from the Red Junglefowl (RJF) about 8000 years ago [1,2], and the changes in morphology, physiology and behaviour as a response to this have been immense. For example, most domesticated chickens grow to at least twice the size of RJF, become sexually mature at a lower age, lay manifold more and larger eggs, show a wide variation in plumage colour and structure, and have a different behaviour in a number of contexts, such as reduced fearfulness [3-6]. In general, domestic animals are assumed to have adapted to a life among humans by evolving higher flexibility in diet, better ability to breed in captivity, less stress susceptibility, and a more socially tolerant disposition [5,6]. It has been suggested that epigenetic mechanisms might be involved in cases like this [7] where wide-encompassing phenotypic changes occur in a short evolutionary time.

However, there is limited knowledge of the extent to which expression and epigenetic profiles are inherited in animals. Reliable inheritance is necessary in order for epigenetic variation to be a major component of any evolutionary process. We have earlier shown that stress-induced modifications in both behaviour and brain gene expression profiles in domestic chickens are to some extent transferred to the offspring [8,9], and other studies have shown similar transgenerational transmission in other species, including humans [10-12]. This indicates that some epigenetic variation may indeed be inherited, but the details and significance of this, as well as its putative evolutionary significance, remain to be elucidated.

One of the possible epigenetic mechanisms, which could be related to variation in gene expression, is methylation of cytosine, preferentially in so called CpG-islands of promoter regions [13,14]. Therefore, we targeted methylation and gene expression simultaneously to investigate whether any of those, or both, would differ between two populations of chickens, recently separated by domestication. We hypothesised that both methylation and gene expression would differ between the populations and show transgenerational stability, opening the possibility for both to be involved in domestication-related phenotypic changes.

By using expression and methylation arrays on hypothalamus samples, we show that profiles of gene expression as well as promoter methylation differ between domesticated White Leghorn layer chickens and their ancestors, the Red Junglefowl. There were also similar differences, although less pronounced, between phenotypically different families within breeds. The differences were largely maintained in the offspring, demonstrating a reliable inheritance of epigenetic states, and for some of the genes the differential methylation was maintained after eight generations of intercross. Our results therefore suggest that selection of favourable epigenetic variants may have been an important aspect of chicken domestication.

RESULTS AND DISCUSSION

Brain Gene Expression Differences within and Between Populations

In this experiment, we studied variations in gene expression and methylation in brains of RJF and domesticated White Leghorns (WL), and their offspring. We focussed on thalamus and hypothalamus, brain regions involved in fear and stress responses, both of which have changed significantly during domestication [3,6]. Within each population, we selected parental animals with divergent phenotypes in order to maximise the within population genetic variation. Specifically, we used two pairs of each population, with pairs within population differing in their behaviour in a series of previously validated tests of stress reactions in chickens [6,15]. From these, totally 73 offspring were hatched and reared until three weeks of age, when they were tested in a fear test, similar to that used in the parents.

In both breeds, body weights differed between families in both generations, and behavioural scores, as measured in the fear tests, differed between families in both generations of WL, but not RJF. Hence, morphological, and to some extent behavioural, phenotypes showed a significant and transgenerationally stable variation in the animals used for the present study. It should be noted that phenotyping was done at different ages in the two generations, which may have been the reason for the lack of transgenerational correlation in fear behaviour in RJF. All eight parents were sacrificed at an age of 373 days, and 48 offspring (12 from each pair) at 21 days, and from each brain, the thalamus-hypothalamus region was removed for extraction of both DNA and mRNA. For the offspring, eight pools of both were prepared, each consisting of six same-sex samples within families. Hence, there were in total eight parental single-animal samples, and eight pools of offspring samples. The mRNA was hybridized to a 38K Affymetrix chicken gene expression microarray, and the DNA was used for subsequent tiling array analysis of methylation. Between populations, there were in total 281 significantly (FDR-corrected $P < 0.05$) differentially expressed (DE) genes in the parents, and 1674 in the offspring. The lower number of DE genes in the parents could possibly be an effect of the lower power of detection given the smaller biological sample size in this generation. Between families within populations, only a few genes were significantly DE, and also DM was less frequent between families. This indicates that expression and methylation profiles are relatively stable within breeds, but both may have changed considerably during domestication.

Transgenerational Stability of Gene Expression Profiles

Out of the significantly DE genes in the parents (comparing populations), 86% percent (n = 242) were also significantly DE in the offspring, and there was a distinct similarity in the expression differences in both generations (Figure 1a). The overall pattern of fold-change levels between populations (regardless of whether they were significant) was strongly correlated over generations (Figure 2 a), further showing a transgenerational stability in gene expression profiles. Also within populations, the overall pattern of fold-change levels between families was highly correlated across generations (Figure 2 b-c).

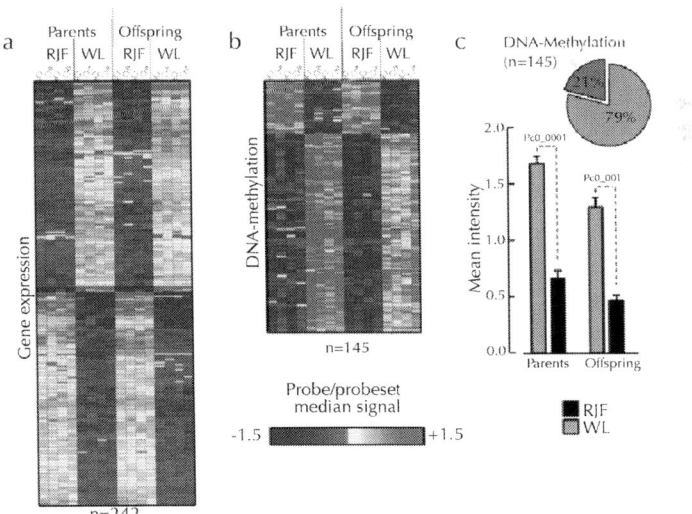

Figure 1: Gene expression and methylation differences between populations and across generations. a. Heat map, showing the clustering of 242 differen-

tially expressed genes, comparing parental Red Junglefowl (RJF) and White Leghorn (WL) layers, and their offspring. b. Heat map, showing clusters of differentially methylated genes comparing parental RJF and WL, and their offspring. Note that the gene set in b is not the same as in a.**c**. Average methylation levels (signal intensity from the microarray) for RJF and WL in the 145 promoter regions included in the heatmap in panel b. Circle-diagram displays the percentage of the promoter regions which were hypermethylated (green) and hypomethylated (red) in White Leghorns.

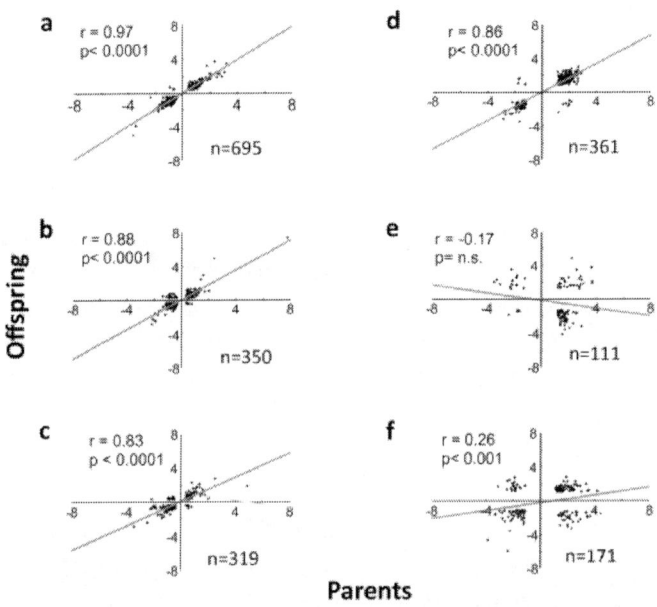

Figure 2: Correlations of differential expression and methylation of genes between generations. a-c, Correlations between generations of differential gene expression, comparing Red Junglefowl and White Leghorn (a); families within Red Junglefowl (b); and families within White Leghorns (c); d-f, Correlation between generations of differential methylation, comparing Red Junglefowl and White Leghorn (d); families within Red Junglefowl(e); and families within White Leghorns (f). The genes included in the analyses were selected based on the fact that they simultaneously occurred in both parents and offspring among top 1000 DE or DM, sorted by fold change.

We further used signalling intensities of individual probesets on each microarray to correlate global expression levels between parents and their own offspring, compared to offspring of other birds, and found a significantly higher correlation within families than between (mean difference in correlation coefficients 0.0017 ± 0.0002 (SEM), $t = 8.2$, $P < 0.001$). This was true both for RJF and WL, and further supports that specific brain gene expression profiles are indeed inherited.

Gene Methylation: Inheritance and Differences between Populations

For analysis of differential methylation (DM), we selected 3623 genes from the list of genes which had the highest fold changes in DE in both generations, both in the between- and within-population comparison. Note that only 281 of these were significant in parents and 1674 in offspring. For each of these genes, 50-75 bp-probes representing a region spanning from -7.25 kb upstream to +3.25 kb downstream of the transcription start point (hence mostly covering promoter regions and other cis-acting regulatory elements) were placed on a custom made tiling array. Methylated DNA immune precipitation (MeDIP) was used to enrich methylated DNA fragments, and after labelling and hybridisation, the relative levels of methylated to un-methylated DNA was assessed for each probe.

Out of the 3623 selected genes, 239 were significantly DM (FDR-corrected $P < 0.05$) when comparing RJF and WL parents, and 821 were DM in the corresponding comparison in the offspring. A smaller number were classified as DM when comparing between families within population (Table S2). A heat map of the genes classified as DM in both generations showed a highly consistent pattern across generations (Figure 1b). Furthermore, DM levels were significantly correlated between generations when comparing RJF with WL (Figure 2 d), and also to a lesser degree when comparing WL, but not RJF families (Figure 2 e-f).

Of the 145 genes which were significantly DM in both generations 79% were hypermethylated in WL (Figure 1c). This is a highly significant bias ($\chi^2 = 49.8$, $P < 0.0001$), indicating that this breed has acquired novel methylation patterns during its selection history.

We further analysed the relationship between DM and DE on the 3623 selected genes. There was no overall correlation between the level of DM of a gene (% of DM probes) and the degree of DE of the same gene. Furthermore, there was no overrepresentation of DE genes among the top 100 DM promoters when compared to a random sample of 100 DM genes ($\chi^2 = 2.1$, P > 0.05). This is contrary to the common notion that methylation causes down-regulation of gene expression, but similar findings have recently been reported from other species, for example humans [16,17]. The finding is quite surprising, and indicates that the specific sites of methylation may be of major importance for gene regulation. For example, there may be a substantial difference between methylation of transcription factors compared to insulator sequences. Since we only analysed a 10 kb region around the transcription start site of each gene, we can not exclude that DM in other, more distant regulatory regions may be more closely connected to the expression level.

To illustrate examples of the transgenerationally stable methylation patterns observed, we show methylation graphs for four genes (*ABHD7, GAB1, KSR1* and *PCDHAC1*) in Figure 3. In all four, the methylation pattern was reliably inherited, shown by the fact that the DM pattern was highly similar in parents and offspring. *ABHD7* showed extensive DM ranging several kb downstream of the transcription start site. In none of the four genes, the significantly DM loci were in CpG-islands, so methylation must have targeted cytosines in other genomic contexts. Extensive methylation of non-CpG regions have recently also been reported for the human methylome [16,17], and it remains unknown which functions these epigenetic variants may serve.

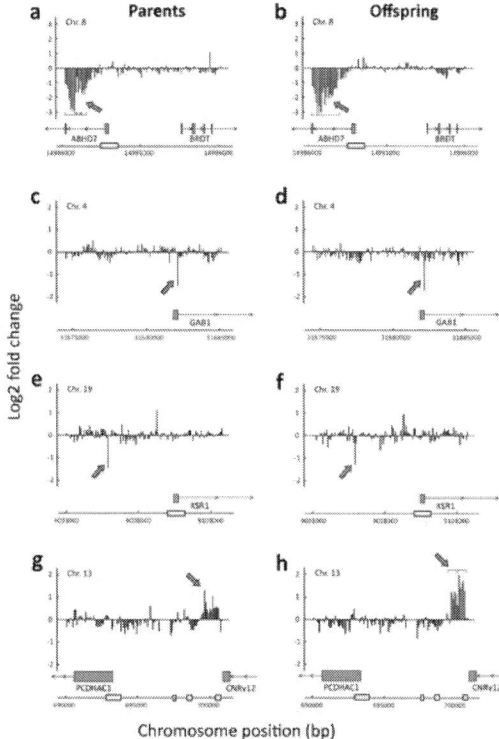

Figure 3: Transgenerational stability of methylation patterns in specific genes. a-h, Differential methylation levels (Log2 fold change) of promoter regions, comparing Red Junglefowl and White Leghorn, are shown with a resolution of 50-75 bp-regions in parents and offspring for each of the genes a, b, *ABHD7*, c, d, *GAB1*, e, f, *KSR1*, and g, h,*RAPGEF1*. Transcription direction and exons (blue boxes) are shown, and red arrows point at bars or groups of bars where significant levels of differential methylation are found (also indicated by red bars). Locations of CpG-islands are indicated in yellow for each region.

Verification of Differential Methylation with Independent Animals, Tissues and Method

To verify the results of the array-based methylation analysis, we arbitrarily selected four genes, which were DM on the tiling arrays in either parents or offspring, *FUCA1, PCDHAC1, TXNDC16,* and*RUFY3*, and replicated the findings for those, using a different technique and a different animal material. Hypothalamus/thalamus regions from eight five-weeks old RJF and eight WL (same strains as earlier, but different parents) were dissected and treated as described above. The DNA was bisulfite-treated, and the degree of methylation was determined in the regions that were significantly DM on the tiling array using methylation sensitive high resolution melting (MS-HRM) analysis.

All four genes were significantly DM in the same direction as found on the tiling array (*FUCA1* and*PCDHAC1* hypermethylated in WL; *RUFY3* and *TXNDC16* hypomethylated) (Figure 4a). This suggests that the tiling array produced reliable results and that the observed methylation differences are representative for the population differences at large.

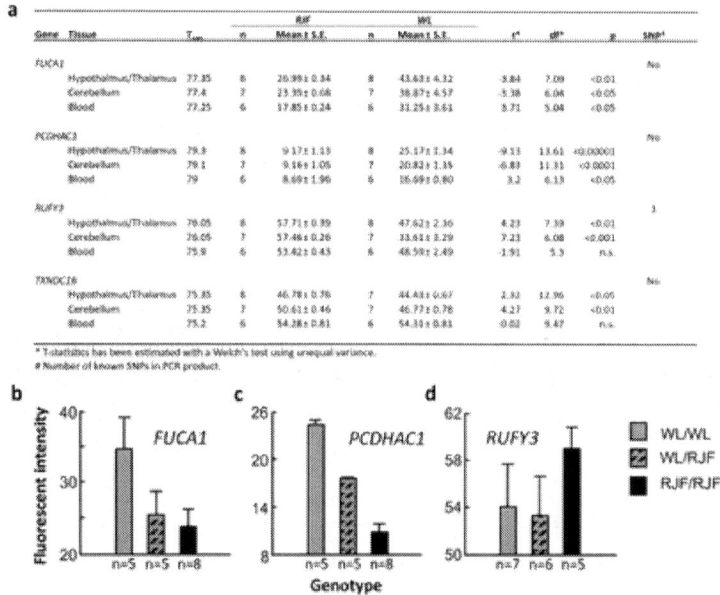

Figure 4: Verification of differential methylation of four arbitrarily chosen genes. a. Methylation differences between Red Junglefowl and White Leghorn in three different tissues, estimated by methylation sensitive high resolution melting (MS-HRM) analysis. The table shows the normalized fluorescent intensities together with the statistical analysis of 6-8 individual samples per population, at a temperature where positive and negative methylation controls showed highest intensity difference (T_{md}). The right column shows the number of SNPs present in the PCR-product amplified by the primers for the

region analysed.b-d, The average methylation level of the promoter regions of three of the genes, *FUCA1*(b), *PCDHAC1* (c) and *RUFY3* (d), in blood samples from F8-generation intercross birds between Red Junglefowl and White Leghorns. The birds differed as indicated in their genotype on an SNP-marker close to the differentially methylated locus.

In order to check for tissue-specificity of the DM, we also performed HRM on the same four genes, using DNA-pools prepared from cerebellum and blood from the offspring samples included in the tiling arrays. All four genes were significantly DM in cerebellum. In blood, *FUCA1* and *PCDHAC1* were significant, while *RUFY3* showed a tendency for DM (P = 0.08) (Figure 4 a). The fact that *TXNDC16* was not DM in blood indicates that this gene shows tissue-specific, heritable methylation.

Genetic Stability of Methylation Differences

There is a risk that the methylation differences detected by the MeDIP technique could be affected by sequence differences in the promoter regions used for the arrays. To exclude this possibility, we used the recently published resequencing data of Red Junglefowl and domestic chickens [18] to check the 145 significantly DM probes in both parents and offspring for possible deletions, insertions and SNP density. Apart from occasional SNPs, no major sequence differences were detected.

The methylation differences observed may be a result of either inheritance of the epigenetic changes independently of genetic changes, or result from sequence differences which secondarily affect methylation at close or remote loci. This is more difficult to differentiate, since it would require extensive resequencing data of the individuals actually used in the study, combined with, for example, methylation QTL-studies.

To suggestively analyse whether differential methylation of specific loci are caused by sequence differences we decided to study its genetic stability and segregation over several generations. For this purpose, we used a total of 18 birds from the eighth generation of an intercross between RJF and WL. In this population, genetic recombinations in each generation have broken up the linkage between adjacent loci, and we could therefore check for both stability of the methylation sites, and for possible cis- or trans-regulation of these.

From this group of advanced intercross birds, we selected individuals, which were homozygous for either the WL or RJF-allele, or heterozygotes, of an SNP located within 176-1449 kb of the locus showing DM. Using HRM analysis on DNA from blood, extracted from these different genotypes, we again analysed the methylation on *FUCA1*, *PCDHAC1* and *RUFY3* in these individuals (Figure 4b). For *FUCA1*, we found two different non-significant, but distinct, methylation levels, where the birds homozygous for the WL-marker were hypermethylated, and heterozygotes were similar to the ones homozygous for the RJF-marker ($P = 0.07$). With respect to *PCDHAC1*, the three genotypes were significantly different ($P < 0.001$), with heterozygotes having a methylation level falling between the hypermethylated WL homozygotes, and the RJF genotypes. *RUFY3* showed a high level of methylation, which was not significantly different between the three genotypes. Hence, two of the three DM loci were stable over the eight generations of intercrossing, and tended to segregate according to genotype at the locus. This is consistent with a cis-regulating mechanism, showing a dominant inheritance of hypomethylation in genotypes with RJF alleles for *FUCA1*, and an intermediate, codominant inheritance in *PCDHAC1*. *RUFY3* may possibly be under control of trans-acting loci, which have segregated during the intercrossing.

Although these results are not conclusive, they suggest that sequence differences may determine the DM for at least two of the three loci, possibly for all of them. This further suggests that selection during domestication may have targeted genotypes which modify the epigenomes, perhaps affecting phenotypes indirectly.

Genetic Pathways

To examine which genetic pathways and functions that may have been affected by DE and DM, we performed a gene ontology (GO) analysis. We analysed the DM and the DE genes in each generation separately, and then selected those GO-terms and KEGG pathways ($P < 0.1$), which were significantly enriched in both generations.

A majority of the enriched GO terms are related to phosphorylation and kinase activity, important aspects of intercellular signalling. Looking specifically at the KEGG pathways enriched among DM and DE genes in offspring only (where the biological sample is considerably larger), the analysis shows that MAPK signalling pathway (which, for example, is associated with stress responses), long-term potentiation (affecting memory consolidation), neurotrophin signalling pathway (involved in neural differentiation) and GnRH signalling pathways (related to reproduction) are enriched. All these are potentially interesting from a domestication perspective, in that they may be related to well documented differences between RJF and WL in stress tolerance, behaviour and reproduction.

Over-Representation of Epigenetically Affected Genes in Selective Sweep Regions

We considered that the epigenetic differences between the layer breed and their ancestor could reflect general effects of selection during domestication, as suggested above, perhaps being related to differences in the domestication induced phenotypes, such as growth, feeding behaviour and social tolerance. If so, we would expect the epigenetic differences to be accumulated in genomic regions which have been under selection during domestication. Therefore, we compared our data to one of our earlier, and recently published, datasets on chickens [18]. This dataset consists of an extensive list of selective sweeps related to chicken domestication, based on resequencing of populations of RJF and a number of domesticated breeds. In total, 149 selective sweeps present in all domestic chickens, and 134 present in egg laying breeds only, were used. A sweep was defined as a 40 kb region where the heterozygosity Z-score was below -4.

There were 216 DE genes (DE in both generations) with annotated loci within the 975 Mb of the genome covered by the sweep analysis. Five of them were situated within 50 kb of selective sweeps present in all domestic chickens (non-significant association, based on a permutation test; $P > 0.1$), and nine in the laying breed sweeps (significantly more than expected by chance; $P < 0.05$).

We performed the same analysis on 134 DM loci, and found that four were within 50 kb of sweeps in all domestic chickens (non-significant; $P > 0.1$), and six in laying breed sweeps ($P < 0.05$). The significant overlapping genes in laying breed sweeps are shown in It is interesting to note that *ABHD7*, which was the strongest DM and one of the strongest DE genes in our experiment, is positioned in a laying breed sweep. This gene is named *EPHX4* in humans, and is related to detoxification of exogenous chemicals [19]. Based on its position in a selective sweep, and its differential methylation and expression, it would appear that the epigenetic variant of the gene (or the genotype affecting the epigenetic state of it) may have been selected during domestication. *KSR1*, an important gene in MAPK/Ras dependent signalling [20], as well as*ADRA2C*, an alpha adrenoreceptor that may be related to egg laying [21] and regulation of the sympathetic stress reaction [22], are also situated in laying breed sweeps.

Although our data do not allow us to conclude on which genes and which sweeps that are associated with specific phenotypes, they suggest that selection of epigenetic variation may have been an important part of chicken domestication.

General Discussion

Our findings show that differential methylation and gene expression in hypothalamus/thalamus are abundant in a comparison between domesticated White Leghorn layers and their wild ancestors, the Red Junglefowl. Many of these epigenetic differences are inherited, demonstrating transgenerational stability. It is possible that these differences are a result of selection during domestication, targeting either sequence differences which affect epigenetic states of specific loci, or epigenetic states which are not related to sequence differences.

The causal relationship between methylation and gene regulation is not clear, since differential methylation was associated with both up- and down-regulation of the gene expression, or did not affect it at all. Since similar dissociation between methylation and gene expression has recently been found in the human genome as well [17,23], this indicates that epigenetic regulation is more complex than previously assumed. Whereas it is often believed that methylation of promoter regions is associated with down-regulation of gene expression, our results indicate that gene regulation is more complex than so. For example, chromatin structure may be more important than commonly assumed. Furthermore, we found that CpG-islands are not always methylated, so there may also be evolutionary contraints on methylation sites, hence affecting the rate with which epigenetic adaptations may

site (Ensemble genebuild WASHUC2), were used to design a custom 385 K DNA-methylation tiling array (Roche-NimbleGen). In total 3623 promoter regions were tiled to the array, with 50-75 mer probes and 100 bp median spacing, by the Madison design team at Roche-NimbleGen.

Methylated DNA Immunoprecipitation (MeDIP)

Protocols of MeDIP with buffer descriptions and general procedures have been published elsewhere[35,36]. Fragmentation of 6 µg thalamus/hypothalamus DNA was performed using a BRANSON sonifier 250 with a 13 mm disruptor horn (101-147-037) and a 3 mm tapered microtip (101-148-062). Samples were diluted with 450 µl 1 × TE in 1.5 ml tubes and sonicated at 10% amplitude by short 0.5 sec pulses (20 in total) with a rest between pulses of 0.5 sec. Fragment lengths of between 300-1000 bp were verified on a Bioanalyzer (Agilent technologies). After sample denaturation 10 min at 95°C, reference samples of 10 µl was taken from each of the original samples and frozen. The remaining samples underwent methylated DNA immunoprecipitation by first diluting them in 1 × TE to 450 µl, adding 51 µl of 10 × IP buffer and 10 µg of 5-meC antibody (Diagenode). Samples were then incubated in 4°C for 2 h on a rotating platform, whereby 50 µl of clean Dynabeads Protein G (Invitrogen) in 1 × IP buffer was added and followed by an identical 2 h incubation. The beads-antibody-antigen complex was washed 3 times by placing the samples on a DynaMag-spin magnet, discarding the supernatant and adding 1 ml of 1 × IP. Complex digestion was done by adding 250 µl of Proteinase K digestion buffer and 5 µl Proteinase K (20 mg/ml), followed by rotation over night in 50°C. DNA was further purified by phenol/chloroform procedures with glycogen/ethanol -80°C precipitation. The pellet were washed in 100% ethanol and resuspended in 60 µl 1 × TE. All samples, references as well as MeDIP›s, were then whole genome amplified using the WGA2 kit (Sigma-aldrich) and purified with QIAquick PCR purification kit (Qiagen).

DNA-Methylation Tiling Array Labeling and Hybridization

Labeling and hybridization was performed at Roche-NimbleGen service lab at Iceland, Reykjavik, using standard protocols [37]. In short, MeDIP and reference samples were labeled with Cy5 and Cy3 respectively, using the NimbleGen Dual-Color DNA Labeling Kit. The MeDIP-Cy5 and reference-Cy3 samples from each tissue sample were then co-hybridised to the DNA-methylation tiling array using the NimbleGen Hybridization Kit and Hybridization System. After washing with NimbleGen Wash Buffer Kit the slides were scanned by a NimbleGen MS 200 Microarray Scanner.

Tiling Array Data Analysis

Methylation data analysis was performed using Bioconductor in the open source R statistical software environment [38]. To not loose genome wide methylation differences, the RINGO package [39] was used to preprocess the data within arrays by Tukey's biweight normalization and between arrays with A-quantile normalization. Significantly differentially methylated probes (FDR adjusted P-values) were extracted using the limma package [32]). Since promoters sometimes involved more than one significant probe, in all comparisons with the gene expression data and correlations across generations, only the most significant probe of each promoter was considered. All significant probes that were stable across generations were checked for the occurrence of SNPs using a list of SNPs detected in a multibreed resequencing study recently published[18].

Methylation Sensitive High Resolution Melting Analysis

Verification of differentially methylated genes, and analysis of differential methylation in alternative tissues and in F8-intercross birds, was done by methylation sensitivehigh resolutions melting (MS-HRM) analysis, principally as described by [40]. If not said otherwise, all procedures followed manufactures recommendations. DNA was prepared from brain tissues as above and from blood using the DNeasy

Blood and Tissue kit adjusted for nucleated blood (Qiagen).

Positive control samples were synthesized by in vitro methylation, using a nuclease-free water diluted reaction mix of 16.5 µl, including an all bird pool of 1 µg DNA, 2 µl 10× NEBuffer2, 2 µl SAM (640 µM), 1 µl SssI methylase (4 U/µl) (New England BioLabs Inc.). After 2 h of incubation at 37°C, an additional 2.5 µl SAM was added to each sample, followed by another 2 h incubation and then termination by heating at 65°C for 20 min.

Negative control samples were synthesized by whole genome amplification on the same all bird DNA pool (10-20 ng/µl) as for the positive control using the REPLI-g Mini Kit (Qiagen). The whole volume of amplified negative controls were then mixed with 200 µl Buffer AL and 200 µl ethanol (99%) and purified with the DNeasy Blood and Tissue kit (Qiagen). 1 µg DNA from both individual samples and controls were bisulfite treated using the EpiTect Bisulfite Kit (Qiagen).

PCR and High resolution melting was performed on a Rotor-Gene 6000 thermocycler (Corbett Research). 1 µl of the bisulfite treated samples/controls were prepared in a 10 µl PCR mix using EpiTect HRM PCR Kit (Qiagen). A calibration series was also amplified using a mixture of positive and negative controls at 100%, 75%, 50%, 25% and 0% of methylated DNA. PCR was performed in 45 cycles as follows: denaturation 10 s at 95°C, annealing at 30 s 54-55°C (primer dependent) and extension 10 s at 72°C. MS-HRM was run in the interval of 70°C to 90°C, with a 2 s 0.05°C steps, acquiring fluorescence data at the Rotor-Gene HRM channel. Primer sequences and annealing temperatures can be seen in All MS-HRM reactions were run in triplicates.

Annotation and GO Analysis

Affy Chicken ID, EntrezGene ID, EnsembleGene ID, WikiGene ID and chromosomal regions were extracted from the Affymetrix annotation file (release 29), and further annotated with Ensemble's BioMart tool [41]. We used DAVID 6.7 http://david.abcc.ncifcrf.gov *webcite* to extract significantly enriched gene ontology terms and KEGG pathways within our datasets [42,43]. To increase the possible DAVID hits we first extracted the homologous human Ensembl ID of our chicken genes in BioMart [41]. CpG island prediction was performed with EMBOSS CpGPlot [44].

Analysis of Sweep Overlaps

219 DE genes and 134 DM promoters (significant in both parents and offspring) fell within the 975 Mb that previously has been searched for selective sweeps[18]. To investigate whether genes or promoters were significantly enriched in sweep regions, 1000 sets of random intervals were generated over the 975 Mb for each analysis, each interval in each set chosen to represent one DE or one DM gene. The number of overlaps between the randomly generated interval and a sweep (within 50 kb of the sweep) was compared to the actual number of real overlaps. A probability of the observed coincidence of less than 5% was taken as a significant association between DM/DE genes and sweeps.

AUTHORS' CONTRIBUTIONS

DN performed the experiments, analysis and lab work, JB and MJ assisted in HRM-analysis, CR performed the sweep analysis, DW performed the F8-breeding and genotyping, LA developed the intercross analysis, and PJ, together with DN, planned and conceptualised the experiment, and wrote the paper. All authors read and approved the final manuscript.

ACKNOWLEDGEMENTS

The project was supported by grants from The Swedish Research Council (2008-14496-59340-36) and The Swedish Research Council for Environment, Agricultural Sciences and Spatial Planning (221-2007-838).

REFERENCES

1.   Fumihito A, Miyake T, Takada M, Shingu R, Endo T, Gojobori T, Kondo N, Ohno S:Monophyletic original and unique dispersal patterns of domestic fowls. *Proceedings of the National Academy of Sciences of the United States of America* 1996,93:6792-6795.

2. Tixier-Boichard M, Bed'hom B, Rognon X: Chicken domestication: From archeology to genomics. *CR Biologies* 2011, in press.

3. Schütz K, Kerje S, Carlborg Ö, Jacobsson L, Andersson L, Jensen P: QTL analysis of a red junglefowl x White Leghorn intercross reveals trade-off in resource allocation between behaviour and production traits. *Behavior genetics* 2002, 32:423-433.

4. Wright D, Kerje S, Lundström K, Babol J, Schutz K, Jensen P, Andersson L: Quantitative Trait Loci Analysis of Egg and Meat Production Traits in a red junglefowl/White Leghorn Cross. *Animal Genetics* 2006, 37:529-534

5. Price EO: *Animal domestication and behavior*. Wallingford: CABI Publishing; 2002.

6. Campler M, Jöngren M, Jensen P: Fearfulness in red junglefowl and domesticated White Leghorn chickens. *Behavioural Processes* 2009, 81:39-43

7. Jablonka E, Raz G: Transgenerational Epigenetic Inheritance: Prevalence, Mechanisms, and Implications for the Study of Heredity and Evolution. *The Quarterly Review of Biology* 2009, 84:131-176

8. Lindqvist C, Jancsak AM, Nätt D, Baranowska I, Lindqvist N, Wichman A, Lundeberg J, Lindberg J, Törjesen PA, Jensen P: Transmission of stress-induced learning impairment and associated brain gene expression from parents to offspring in chickens. *PLoS ONE* 2007, 2:e364

9. Nätt D, Lindqvist N, Stranneheim H, Lundeberg J, Torjesen PA, Jensen P: Inheritance of Acquired Behaviour Adaptations and Brain Gene Expression in Chickens. *PLoS ONE* 2009, 4:e6405.

10. Franklin TB, Russig H, Weiss IC, Gräff J, Linder N, Michalon A, Vizi S, Mansuy IM:Epigenetic Transmission of the Impact of Early Stress Across Generations. *Biological Psychiatry* 2010, 68:408-415

11. Anway M, Cupp AS, Uzumcu M, Skinner MK: Epigenetic trans-generational actions of endocrine disrupters and male fertility. *Science* 2005, 308:1466-1469

12. Kaati G, Bygren LO, Pembrey M, Sjostrom M: Transgenerational response to nutrition, early life circumstances and longevity. *Eur J Hum Genet* 2007, 15:784-790

13. Richards EJ: Inherited epigenetic variation - revisiting soft inheritance. *Nature Reviews Genetics* 2006, 7:395-402

14. Zhang T-Y, Meaney MJ: Epigenetics and the Environmental Regulation of the Genome and Its Function. *Annual Review of Psychology* 2010, 61:439-466. Forkman B, Boissy A, Meunier-Salaün M-C, Canali E, Jones RB: A critical review of fear tests used on cattle, pigs, sheep, poultry and horses.*Physiology & Behavior* 2007, 92:340-374.

15. Lister R, Pelizzola M, Dowen RH, Hawkins RD, Hon G, Tonti-Filippini J, Nery JR, Lee L, Ye Z, Ngo Q-M, *et al.*: Human DNA methylomes at base resolution show widespread epigenomic differences. *Nature* 2009, 462:315-322

16. Weber M, Hellmann I, Stadler MB, Ramos L, Pääbo S, Rebhan M, Schuberler D:Distribution, silencing potential and evolutionary impact of promoter DNA methylation in the human genome. *Nature Genetics* 2007, 39:457-466

17. Rubin C-J, Zody MC, Eriksson J, Meadows JRS, Sherwood E, Webster MT, Jiang L, Ingman M, Sharpe T, Ka S, *et al.*: Whole genome resequencing reveals loci under selection during chicken domestication. *Nature* 2010, 464:587-593.

18. Bagryantseva Y, Novotna B, Rossner PJ, Chvatalova I, Milcoca A, Svecova C, Lnenickova Z, Solansky I, Sram RJ: Oxidative damage to biological macromolecules in Prague bus drivers and garagemen: impact of air pollution and genetic polymorphisms. *Toxicology Letters* 2010, 199:60-68.

19. Fusello AM, Mandik-Nayak L, Shih F, Lewis RE, Allen PM, Shaw AS: The MAPK scaffold kinase suppressor of Ras is involved in ERK activation by stress and proinflammatory cytokines and induction of arthritis. *Journal of Immunology* 2006, 177:6152-6158.

20. Moudgal RP, Razdan MN: Induction of ovulation in vitro by LH and catecholamines in hens is mediated by alpha-adrenergic receptors. *Nature* 1981, 293:738-739

21. Puvadolpirod S, Thaxton JP: Model of physiological stress in chickens 2. Dosimetry of adrenocorticotropin. *Poultry Science* 2000, 79:370-376.

22. Gibbs JR, van der Brug MP, Hernandez DG, Traynor BJ, Nalls MA, Lai S-L, Arepalli S, Dillman A, Rafferty IP, Troncoso J, et al.: Abundant Quantitative Trait Loci Exist for DNA Methylation and Gene Expression in Human Brain. PLoS Genet 2010, 6:e1000952

23. Sharma A, Singh P: Detection of Transgenerational Spermatogenic Inheritance of Adult Male Acquired CNS Gene Expression Characteristics Using a DrosophilaSystems Model. PLoS ONE 2009, 4:e5763

24. Zhang X, Ho S-M: Epigenetics meets endocrinology. Journal of Molecular Endocrinology 2011, 46:R11-R32

25. Miller D, Brinkworth M, Iles D: Paternal DNA packaging in spermatozoa: more than the sum of its parts? DNA, histones, protamines and epigenetics. Reproduction 2010, 139:287-301.

26. Molinier J, Ries G, Zipfel C, Hohn B: Transgeneration memory of stress in plants. Nature 2006, 442:1046-1049

27. Zhai J, Liu J, Liu B, Li P, Meyers BC, Chen X, Cao X: Small RNA-Directed Epigenetic Natural Variation in Arabidopsis thaliana. PLoS Genetics 2008, 4:e1000056.

28. Franklin TB, Mansuy IM: Epigenetic inheritance in mammals: Evidence for the impact of adverse environmental effects. Neurobiology of Disease 2010, 39:61-65

29. Schütz K, Kerje S, Jacobsson L, Forkman B, Carlborg Ö, Andersson L, Jensen P: Major growth QTLs in fowl are related to fearful behavior: Possible genetic links between fear responses and production traits in a red junglefowl x White Leghorn intercross. Behavior genetics 2004, 34:121-130

30. Gautier L, Cope L, Bolstad BM, Irizarry RA: Affy - analysis of Affymetrix GeneChip data at the probe level. Bioinformatics 2004, 20:307-315

31. Smyth GK: Linear models and empirical bayes methods for assessing differential expression in microarray experiments. Stat Appl Genet Mol Biol 3 2004. Article 3

32. Hochberg Y, Benjamini Y: More powerful procedures for multiple significance testing. Stat Med 1990, 9:811-818

33. Sturn A, Quackenbush J, Trajanoski Z: Genesis: cluster analysis of microarray data. Bioinformatics 2002, 18:207-208

34.   NimbleGen Systems I: Sample Preparation for DNA Methylation Microarray v1.0. 2007.

35.   Weng Y-I, Huang TH-M, Yan PS: Methylated DNA immunoprecipitation and microarray-based analysis: detection of DNA methylation in breast cancer cell lines. *Methods in Molecular Biology* 2010, 590:165-176.

36.   Roche NimbleGen I: NimbleGen Arrays User's Guide: DNA Methylation Analysis, v7.0. 2009.

37.   Gentleman R, Carey V, Bates D, Bolstad B, Dettling M, Dudoit S, Ellis B, Gautier L, Ge Y, Gentry J, *et al.*: Bioconductor: open software development for computational biology and bioinformatics. *Genome Biology* 2004, 5:R80.

38.   Toedling J, Sklyar O, Huber W: Ringo - an R/Bioconductor package for analyzing ChIP-chip readouts. *BMC Bioinformatics* 2007, 8:221.

39.   Wojdacz TK, Dobrovic A, Hansen LL: Methylation-sensitive high-resolution melting. *Nat Protocols* 2008, 3:1903-1908

40.   Spudich G, Fernandez-Suarez X: Touring Ensembl: A practical guide to genome browsing. *BMC genomics* 2010, 11:295.

41.   Huang DW, Sherman BT, Lempicki RA: Systematic and integrative analysis of large gene lists using DAVID bioinformatics resources. *Nat Protocols* 2008, 4:44-57.

42.   Kanehisa M, Goto S, Furumichi M, Tanabe M, Hirakawa M: KEGG for representation and analysis of molecular networks involving diseases and drugs. *Nucleic acids research* 2010, 38:D355-D360.

43.   Rice P, Longden I, Bleasby A: EMBOSS: The European Molecular Biology Open Software Suite. *Trends in Genetics* 2000, 16:276-277.

Elevated Arousal at Time of Decision-Making is not the Arbiter of Risk Avoidance in Chickens

A. C. Davies[1], A. N. Radford[2], I. C. Pettersson[1], F. P. Yang[3] & C. J. Nicol[1]

[1]Animal Welfare and Behaviour Group, School of Clinical Veterinary Science, University of Bristol, Langford House, Langford, Bristol, BS40 5DU, UK

[2]School of Biological Sciences, University of Bristol, Life Sciences Building, 24 Tyndall Avenue, Bristol, BS8 1TQ, UK

[3]College of Bioscience and Biotechnology, Yangzhou University, Yangzhou, Jiangsu, China, 225009

ABSTACT

The somatic marker hypothesis proposes that humans recall previously experienced physiological responses to aid decision-making under uncertainty. However, little is known about the mechanisms used by non-human animals to integrate risk perception with predicted gains and losses. We monitored the behaviour and physiology of chickens when the choice between a high-gain (large food quantity), high-risk (1 in 4 probability of receiving an air-puff) option (HGRAP) or a low-gain (small food quantity), no-risk (of an air-puff) (LGNAP) option.

We assessed when arousal increased by considering different stages of the decision-making process (baseline, viewing, anticipation, reward periods) and investigated whether autonomic responses influenced choice outcome both immediately and in the subsequent trial. Chickens were faster to choose and their heart-rate significantly increased between the viewing and anticipation (post-decision, pre-outcome) periods when selecting the HGRAP option. This suggests that they responded physiologically to the impending risk. Additionally, arousal was greater following a HGRAP choice that resulted in an air-puff, but this did not deter chickens from subsequently choosing HGRAP. In contrast to human studies, we did not find evidence that somatic markers were activated during the viewing period, suggesting that arousal is not a good measure of avoidance in non-human animals.

Subject terms:

INNTRODUCTION

An important aspect of adaptive decision-making is the ability to use prior experiences to assess potential gains and losses, thus making predictions about optimal choices[1]. Under natural conditions, trade-offs between high-gain, high-risk options and low-gain, low-risk options often exist and individuals have been shown to alter their behaviour in response to a wide variety of risks, including predation[2], variation in food quality[3], environmental temperature[4] and flight collisions[5]. However, despite the importance of risk assessment in adaptive decision-making, we know relatively little about the mechanisms used by non-human animals to integrate their perceptions of risk with likely gains and losses.

Decisions must frequently be made rapidly with imperfect knowledge of the available options, making it impossible for an individual to "weigh-up" the costs and benefits of decision outcomes accurately[6]. Under such conditions, alternative decision-making strategies may be employed[7, 8]. It has been suggested, for example, that when making decisions in which the outcome is uncertain, humans rely more on emotional than conscious thought processes[9], to provide a rapid but crude appraisal of the available options[10]. This theory, known as the somatic marker hypothesis[9], proposes that physiological responses which have previously been associated with

the available options (stored as "somatic markers") are recalled during this assessment period to aid decision-making. In humans, this process occurs before an individual is consciously aware of the "advantageous" or "disadvantageous" options[11].

The neural mechanisms underlying decision-making under uncertainty have been investigated in humans to some extent (see ref. 12). A considerable effort has also been made to link these processes with measures of autonomic arousal (e.g. refs. 13, 14, 15), both in anticipation of and as a consequence of decision-making[16]. Although some work has begun to monitor neural processes during decision-making requiring risk assessment in non-human species (see review[17]), it is not known whether other animals also adopt rapid appraisal mechanisms, such as recalling arousal, when assessing risk. In a previous study, we found that behavioural and physiological measures of arousal were detectable when chickens made simple foraging decisions[18], but it was not clear whether arousal influenced choice behaviour or whether it was simply a conditioned anticipatory response to food. In the current study we therefore aimed to monitor behaviour and physiology at different stages of decision-making, when individuals chose between a high-gain, high-risk option and a low-gain, no-risk option.

In the aforementioned human studies, risk perception was monitored by taking measures of sympathetic autonomic arousal (skin conductance reactivity and heart-rate (HR)) during a period in which subjects "pondered" the available options, before they had explicit knowledge of which was the profitable outcome[11, 16]. Although the effect of risk perception on arousal has not yet been studied in chickens, increased arousal has been measured in response to a mild air-puff, which individuals also learned to avoid (Edgar et al. in prep). In the current study we therefore used a high-gain (HG) reward paired with the risk of an air-puff (RAP) as an aversive stimulus, to determine whether the perception of risk increased arousal during decision-making and how it influenced choice behaviour.

During a set of tests, individual hens were given a choice between the HGRAP option and one with a low-gain and no air-puff risk (LGNAP). We monitored indicators of sympathetic autonomic arousal (HR), heart-rate variability (HRV), peripheral body temperature, which are sensitive markers of reward anticipation[18, 19] and aversive

situations[20, 21]. These physiological measures, along with behavioural indicators of arousal (head movements, latency to choose) [18, 22, 23], were taken during a viewing period when chickens were presented simultaneously with the HGRAP and LGNAP options. In addition, we took repeated measures of sympathetic autonomic arousal during a post-decision, pre-outcome (anticipation) period and again when chickens accessed their chosen outcome. This sequential monitoring enabled us to (i) identify precisely when arousal occurred in relation to decision-making and (ii) assess whether autonomic responses (and potentially their associated emotions) influenced choice outcome both immediately and in the subsequent trial.

METHODS

Ethics Statement

All work was conducted under UK Home Office licence (30/2779) and had University ethical approval. We also conducted the study in compliance with ASAB ethical guidelines. The hens were rehomed to small responsible free-range holdings after the study.

Experimental Set-Up and Habituation

Twenty-eight Columbian Black Tail laying-hens were obtained at 17 weeks of age and were group-housed in a single room (see supplementary information for full details).

The experimental room, separated from the home room by a solid metal door and a corridor to prevent noise transfer, contained a T-maze apparatus. In brief, the T-maze consisted of two pens joined by a Perspex tunnel, which included four pulley-operated doors; a start-box, with removal wooden side panels and a pulley-operated entrance door was attached to the tunnel. In each pen, a feeder was fixed to the back wall; an air puff could be delivered when required through this feeder. See supplementary information for full details.

Habituation to HR monitoring and the T-Maze were carried out in parallel (see supplementary information). Throughout the habituation and training phase, hen progress was assessed using training criteria.

Colour-Cue Training: Pre Air-Puff Phase

Twenty-two hens satisfied all habituation criteria and were grouped according to how quickly they were ready for testing (into six groups of 2–4 hens). The day before testing each group of hens began, two pieces of A4 card were stuck to the back of the tunnel (red on left, blue on right) and hens were trained to associate the colour and side of the tunnel with receiving either 1 (low-gain: LG) or 4 (high-gain: HG) pieces of sweetcorn in the feeder. The colour associated with each quantity of sweetcorn was systematically varied so that within each testing group an equal number of hens received four pieces of sweetcorn on the right and left. When the colours had been introduced, each hen was given a set of nine trials (six unidirectional and three free) to strengthen the association. The latency to choose was recorded, as was the choice made during the free trials. If any hen chose the lower quantity side of the T-Maze more than once in the three free trials, additional training trials were conducted.

On the day after each group of hens had been individually trained to recognise the colour-cues, we assessed whether hen behaviour or physiology was affected by food quantity (before we introduced the air-puff) during a further set of 10 training trials. We food deprived the birds for 2 h and put on the ECG pads and HR monitor 15 min before training to allow the hen time to adapt to walking. The first two trials were unidirectional (one to either side of the T-Maze as a "reminder" of the colour cues), followed by eight free trials, to assess whether they remembered which colour was associated with HG. During each of the 10 trials, baseline physiological measures were taken for 10 s after the hen was first placed into the start-box (baseline period). The side panels of the start-box were then removed to reveal the coloured card on the inside of the Perspex tunnel and the individual was confined for a further 10 s (the viewing period). The tunnel door was then raised allowing the hen to enter the tunnel and make a choice by moving towards either of the pen doors. Once a choice had been made the relevant middle door was closed and the hen was confined between the middle and pen doors for 10 s (anticipation period) prior to the pen

door being opened, allowing access to the feeder. Once the hen had reached the feeder, measures were taken during the first 10 s (reward period). The hen was kept within the pen after making a choice for 1 min, before the next test commenced. If a hen failed to leave the start-box after 60 s, she was gently encouraged to move into the tunnel. Similarly if she failed to enter the pen once the pen door had been removed for 120 s, she was gently encouraged into the pen.

Criteria were set (that hen's chose HG at least 70% of the time) to progress to the testing phase. Although all hens showed a greater preference for the HG side, a few hens didn't quite reach criteria, so additional trials were given as necessary.

Testing Phase

Immediately after each individual completed the 10 pre air-puff trials, during the same session, each hen was given a set of 30 free-choice trials in which the HG side of the T-maze became the HGRAP option and the LG side became the LGNAP option. This was achieved by pairing HG with the risk of receiving a single air-puff at the feeder, at a fixed probability of 1 in 4. The air-puff schedule was predetermined for each hen, although criteria were set to prevent more than four air-puffs being delivered consecutively. The air-puff was delivered from outside the pen when the experimenter viewed the hen's head in the feeder. All other aspects of each trial were as described for the pre air-puff phase. If a hen made 10 consecutive choices to the same side, a unidirectional trial was given to the opposite side. During each trial, behaviour and physiology were recorded.

Behavioural and Physiological Measures

The latency to reach the middle door (i.e. time from tunnel door being raised to the start of the anticipation period) and to the feeder (i.e. time from pen door being raised to reaching the feeder) were both recorded using a stop-watch. A CCTV camera was fixed above the start-box and video was continuously recorded using WebCCTV software. The number of head movements made during the viewing period was subsequently recorded using Windows Media Player. The decision outcome was also noted for each trial.

ECG was monitored as in[18] using non-invasive remote telemetric units[24] and cables contained within a harness. The monitor communicated with a base unit (attached to a computer via USB connection) and was controlled using RVC Telemetry Software version 1.5. Measures of HR and HRV were extracted using Spike 2 Software (version 6) from four 10 s periods: baseline, viewing, anticipation and reward. From each 10 s period, an average of HR (bpm) and two measures of HRV – the root mean square of the successive differences between beats (RMSSD) and the coefficient of variance (standard deviation of the mean interval between beats divided by the mean interval between beats – SDNN/RR) – were taken. Baseline measures were taken to control for individual differences at the start of the test. The percentage change in HR between the baseline and viewing periods, between the viewing and anticipation periods and between the anticipation and the reward periods were subsequently calculated and were analysed in addition to absolute values during each period.

Surface body temperature was recorded whilst hens were in the start-box using a thermal video camera (FLIR SC305). Eye and maximum head temperature data were extracted from a clear image during the baseline and viewing periods of each test using FLIR ResearchIR Software version 1.2 SP2. The percentage change in both eye and head temperature between the two periods was subsequently calculated and were analysed in addition to absolute values during the baseline and viewing periods.

Statistical Analysis

Data were analysed using *IBM SPSS Statistics 21*. For each set of data, the assumptions of parametric testing were checked and data were transformed if possible, then analysed using paired-samples t-tests. Where transformations were not possible or unsatisfactory, Wilcoxon tests were used. Because multiple t-tests were conducted on behavioural and physiological data collected from the same hens during the same testing periods, a Benjamini-Hochberg correction was applied to relevant *p*-values from each testing period. There were occasional missing data when HR or temperature could not be obtained. Unless otherwise stated, means ± SE are presented throughout and a measure of effect size is given alongside significant results.

Twenty-one hens completed both the pre air-puff training and the testing phases of the experiment. The pre air-puff data were analysed to check for differences in behavioural and physiological measures of arousal, relating to food quantity (prior to air-puff introduction). An average was taken for each hen from trials when they chose HG and LG. Three hens chose only the HG option, so were excluded from these analyses. Bivariate correlations were also used to identify relationships between HG preference (proportion of times chosen) and individual hen's physiological responses during tests. The data were also analysed to monitor the effect of the choice in one trial on behavioural and physiological measures during the baseline and viewing periods in the subsequent test.

Data from the testing phase were analysed in the same way as the pre air-puff training data. Data were included only after hens had experienced their first air-puff. For each behavioural and physiological measure an average was taken for each hen from tests in which they chose HGRAP and from tests in which they chose LGNAP. We also checked for differences in baseline and viewing period measures in those trials that immediately *followed* an HGRAP choice that had resulted in an air-puff, to identify whether this influenced arousal. Most hens didn't experience more than three or four air-puffs, but a few hens experienced up to nine. Five hens continuously chose HGRAP throughout the testing phase, hence they were excluded from this part of the analyses. Additionally, bivariate correlations were used to identify relationships between HGRAP preference (proportion of times chosen) and individual hen's physiological responses during tests.

RESULTS

Pre Air-Puff Phase

During the pre-air-puff phase, individual hens chose the HG option on average $82 \pm 2\%$ (range: 70–100%) of the time. There were no significant differences in any of the indicators of arousal (latency to choose or to feeder, head movements, HR, HRV and temperature) during trials when hens chose the HG compared with the LG option (paired samples t-tests: all $p > 0.05$). However, the proportion of times hens chose the

HG option during the pre-air-puff phase was significantly, positively correlated with the HR change between the viewing and anticipation periods (Pearson correlation coefficient = 0.45, p = 0.040, n = 21) and was negatively correlated with the change between the waiting and reward periods (Pearson correlation coefficient = −0.49, p = 0.025, n = 21). All other physiological variables showed no significant correlation with the proportion of times HG was chosen.

There were no significant differences in physiological and behavioural measures (HR, temperature, HR and temperature change, head movements) during the baseline or viewing periods in the trial following an HG or LG choice outcome in the previous trial (paired samples t-tests: all p > 0.05).

Air-Puff Testing Phase

During the testing phase, individual hens chose the HGRAP option on average 64 ± 7% (range: 12–100%) of the time.

Were there any significant differences in arousal during trials when hens chose the HGRAP compared with the LGNAP option?

The majority of physiological and behavioural measures taken during the baseline, viewing anticipation and pen periods (HR, temperature, HR and temperature change, head movements) did not differ significantly between trials when hens chose HGRAP compared with LGNAP (supplementary table 1). However, significantly lower latencies to choose (paired samples t-test: t_{15} = 2.32, p = 0.035, eta squared = 0.26, figure 1a), a greater HR change between the viewing and anticipation period (t_{15} = 2.85, p = 0.012, eta squared = 0.35, figure 1b) and a significantly higher reward period HR (t_{15} = 2.80, p = 0.039, eta squared = 0.35, figure 1c) were apparent when hens chose the HGRAP option.

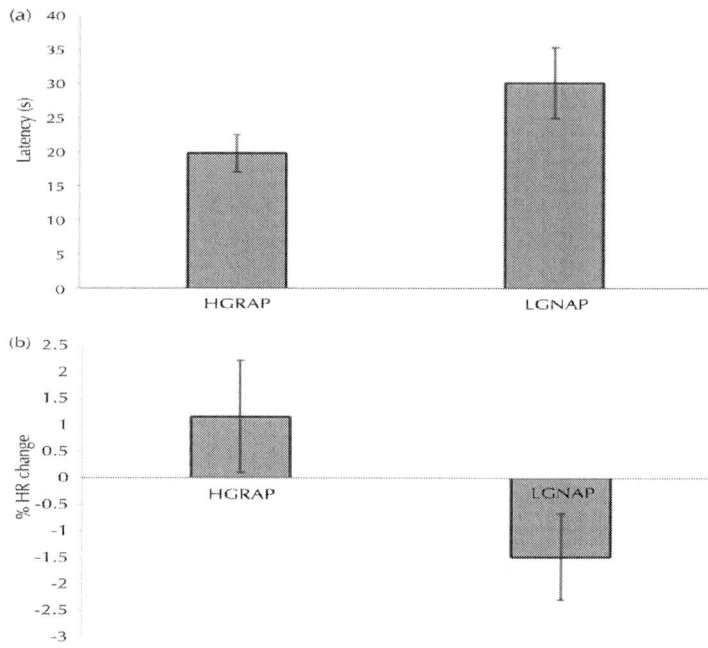

Figure 1: Mean ±1 SE (a) latency to anticipation period (n = 16, p = 0.035) and (b) percentage change in HR between the viewing and anticipation periods (n = 16, p = 0.012) when choosing the HGRAP and LGNAP options.

Were there significant correlations between the proportion of times HGRAP was chosen and the physiological responses during tests?

There were significant, positive correlations between the proportion of times HGRAP was chosen after the first air-puff had been received and the HR during the baseline (Pearson correlation coefficient = 0.61, p = 0.003, n = 21), viewing (Pearson correlation coefficient = 0.67, p = 0.001, n = 21), anticipation (Pearson correlation coefficient = 0.60, p = 0.004, n = 21) and reward (Pearson correlation coefficient = 0.57, p = 0.007, n = 21) periods. There was also a significant negative correlation between the proportion of times HGRAP was chosen and SDNN/RR during the viewing period (Pearson correlation coefficient = −0.49, p = 0.028, n = 20). All other physiological variables showed no significant correlation with the proportion of times HGRAP was chosen.

Were there any Significant Differences in Arousal at the Start of the Subsequent Trial Following a HGRAP or LGNAP outcome?

The majority of dependent variables (HR, temperature, HR and temperature change, latency, head movements) were not significantly different during the baseline and viewing periods of the next trial after hens had chosen LGNAP or HGRAP in the previous trial (supplementary table 2).

Were there any Significant Differences in Arousal at the Start of the Subsequent Trial Following an Air-Puff Outcome?

There were no significant differences in absolute physiological and behavioural measures (HR, temperature, head movements), or changes between baseline and viewing periods (HR and temperature), in the trials after hens received an air-puff compared with trials after a HGRAP choice that resulted in no air-puff being received (supplementary table 3).

Did the Air-Puff Influence Subsequent Choice Behaviour and Arousal?

After an air-puff was received, there was no significant difference in the likelihood of HGRAP being chosen in the subsequent trial compared to when no air-puff was received (HGRAP (after air-puff): 57 ± 9%; HGRAP (after no air-puff): 65 ± 7%; paired samples t-test: $t_{20} = 1.49$, $p = 0.153$). No measures were significantly different when comparing choices of the HGRAP and LGNAP option in the trial following an air-puff (all $p > 0.05$). However, when hens had received an air-puff and chose HGRAP in their subsequent trial, their baseline HR in that subsequent trial was significantly higher than when no air-puff was received in the previous trial and HGRAP was chosen in the subsequent trial ($t_{15} = 2.99$, $p = 0.045$, eta squared = 0.37, figure 2a). Both the maximum head ($t_{14} = 3.15$, $p = 0.035$, eta squared = 0.39, figure 2b) and eye temperature ($t_{14} = 2.84$, $p = 0.033$, eta squared = 0.37, figure 2c) were significantly lower during the subsequent viewing period when HGRAP was chosen following an air-puff compared with no air-puff.

DISCUSSIONS

The role of physiological arousal or stress in decision-making is often invoked but has been little investigated in animals. By partitioning the decision-making process, and monitoring physiology and behaviour at each stage, we found that hens reacted to risk with elevations in physiological arousal, but this did not deter them from choosing high-gain outcomes associated with risk.

It was the perception of risk rather than reward that produced an elevation in HR during the anticipation period preceding each HGRAP choice. During the pre-air puff phase HR was not affected by reward magnitude, but during the testing phase arousal increased between the viewing and anticipation periods when hens chose HGRAP rather than LGNAP. There was no further significant increase between anticipation and reward periods. The only methodological difference between the testing phase and the pre air-puff phase was the risk of an air-puff. Although it has previously been shown that arousal increases in anticipation of conditioned appetitive and aversive events delivered with 100% contingency in chickens (e.g. refs. 18, 19, 20, 21, 22), our work shows that hens are sufficiently sensitive to anticipate a 25% risk of an aversive event. During the anticipation period, hens were therefore responding physiologically to their perception of the risk associated with the HG option.

In addition, several of our measures of arousal were influenced by the nature of the previous decision. During the testing phase, HR was significantly higher during the baseline period, and the maximum head and eye temperature during the viewing period were also significantly lower when HGRAP was chosen in a subsequent trial after an air-puff compared to when it was chosen after no air-puff. It is likely, therefore, that in the subsequent trial we were measuring a residual effect of the increased arousal caused by hens "coping" with the previous HGRAP outcome (e.g. ref. 25). These combined results suggest that receiving an air-puff has a measurable impact on physiological arousal in the subsequent trial.

One of our main findings was that none of the measures of arousal that were associated with the HGRAP outcome affected the birds' subsequent tendency to choose this same outcome. In short, arousal was not a good marker of avoidance. Although, in general, chickens seemed to reduce their preference for the HG option when it was paired with the risk of an air-puff, they showed no decreased probability of choosing it directly after receiving an air-puff. This was despite increased arousal being evident at the time when a choice was being made. Additionally, the proportion of times HGRAP was chosen during the testing phase was significantly, positively correlated with HR during all phases of the test, suggesting that more persistent birds did find the air-puff arousing. It would seem, therefore, that the function of arousal was not in mediating decision-making, but was likely associated with both reward activation and punishment-avoidance systems. Arousal may prepare an individual for fight or flight (e.g. ref. 26) and is not necessarily negative. In this case, the birds that continued to choose the HGRAP option clearly experienced it as positively valenced (e.g. ref. 27). We do not know how individuals made the assessment to reduce their preference for the HGRAP option, but possibly high arousal is recalled during later assessments and is used in conjunction with other information to produce the overall decline in visits.

This analysis leads to an interesting comparison with the somatic marker hypothesis, which proposes that arousal is generated and used (subconsciously) in humans to indicate options that should be avoided[9, 11]. It seems from our work that arousal generated during decision-making is not always used as a marker signifying that an option should be avoided. One possibility is that arousal must be accompanied by some other assessment and coding of long-term gain or loss for it to influence decision-making. A key difference between the Iowa Gambling Task (IGT), which is used to test human decision-making, and the task we developed here is that the IGT produces immediate gains and losses, which lead to long-term advantageous and disadvantageous options[28]. Anticipatory responses are generated prior to choosing the disadvantageous option which humans ultimately learn to avoid (e.g. ref. 11). The options that we presented had no clear advantageous or disadvantageous long-term consequences and although some individuals showed a preference for HGRAP or LGNAP, these individual preferences likely depended on how well each individual coped with the aversive air-puff stimulus. Possibly arousal must be mediated by some neural coding of loss before it is used as a marker that an outcome should be avoided.

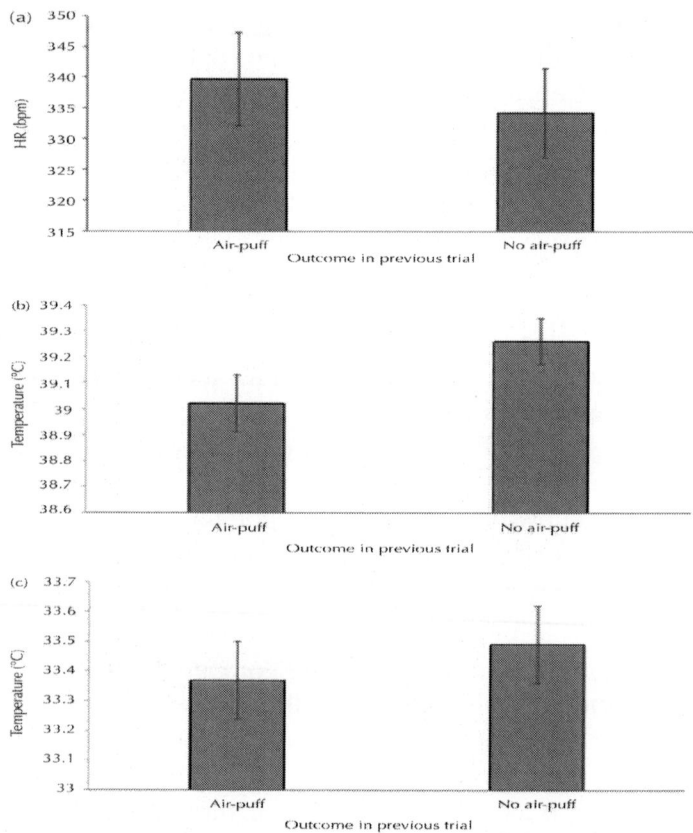

Figure 2: Mean ±1 SE (a) baseline HR (n = 16, p = 0.045), (b) maximum head temperature during the viewing period (n = 15, p = 0.035), and (c) eye temperature during the viewing period (n = 15, p = 0.033) when choosing the HGRAP option following HGRAP outcomes that did and did not result in an air-puff.

Just as arousal does not always lead to avoidance, not all types of 'difficult' decision lead to arousal. For example, arousal is no greater when hens have to choose between finely balanced (options of equal net value) than when they have to choose between options of unequal net value[29]. We suggest that a broad range of decision-tasks should be used before general conclusions can be drawn about the role of physiology in decision-making.

As an additional point, we noted that hens were significantly quicker to make their choice when choosing HGRAP over LGNAP. Heightened arousal (HR) when choosing HGRAP could have resulted in hens approaching the door more quickly or they may have taken less time to consider a HGRAP choice outcome than LGNAP. Alternatively, hens might have been very committed to choosing the HGRAP option and therefore decided more quickly. It has previously been found that chickens with strong and consistent preferences had shorter latencies to choose[30].

In summary, we have found increased arousal in anticipation of the risk associated with the high-gain option. We also found that receiving an air-puff resulted in increased arousal during the subsequent trial, but only when the HGRAP option was chosen again. Interestingly, however, high arousal did not result in hens avoiding the HGRAP option, suggesting that it is not a good measure of avoidance. Although we found that some measures of arousal increased after hens had received an air-puff, our results suggest that this was likely to be a residual effect of the air-puff rather than arousal (somatic markers) mediating subsequent decision-making as seen in human decision-making tasks. We suggest that for elicitation of somatic markers, choice outcomes must be differentiable in long-term gain or loss.

ACKNOWLEDGEMENTS

We would like to thank the BBSRC and UFAW for funding the work and the University of Bristol animal services unit for taking good care of the hens during the study. We would also like to thank Professor Linda Keeling for her helpful comments on the experimental design.

AUTHER CONTRIBUTIONS

A.D., A.R. and C.N. wrote the main manuscript text. A.D., A.R. and C.N. designed the experiment. A.D. and I.P. performed the experiments. A.D. and F.Y. extracted the data and A.D. analysed the data. All authors reviewed the manuscript.

REFERENCES

1.   Weller, J. A., Levin, I. P. & Bechara, A. Do individual differences in Iowa Gambling Task performance predict adaptive decision making for risky gains and losses? *J. Clin. Exp. Neuropsychol.* 32, 141–150 (2010).

2.   Lima, S. L. & Dill, L. M. Behavioral decisions made under the risk of predation: a review and prospectus. *Can. J. Zool.* 68, 619–640 (1990).

3.   Caraco, T. Energy budgets, risk and foraging preferences in dark-eyed juncos (Junco hyemalis). *Behav. Ecol. Sociobiol.* 8, 213–217 (1981).

4.   Cerda, X., Retana, J. & Cros, S. Critical thermal limits in Mediterranean ant species: trade-off between mortality risk and foraging performance. *Funct. Ecol.* 12, 45–55 (1998).

5.   Cuthill, I. & Guilford, T. Perceived risk and obstacle avoidance in flying birds. *Anim. Behav.*40, 188–190 (1990).

6.   McFarland, D. J. Decision making in animals. *Nature.* 269, 15–21 (1977).

7.   Trimmer, P. C. *et al.* Mammalian choices: combining fast-but-inaccurate and slow-but-accurate decision-making systems. *Proc. Biol. Sci.* 275, 2353–2361 (2008).

8.   Chittka, L., Skorupski, P. & Raine, N. E. Speed-accuracy tradeoffs in animal decision making. *Trends. Ecol. Evol.* 24, 400–407 (2009).

9.   Damasio, A. R., Everitt, B. J. & Bishop, D. The somatic marker hypothesis and the possible functions of the prefrontal cortex. *Philos. Trans. R. Soc. Lond. B. Biol. Sci.* 351, 1413–1420 (1996).

10.   Loewenstein, G. F., Weber, E. U., Hsee, C. K. & Welch, N. Risk as feelings. *Psychol. Bull.*127, 267–286 (2001).

11.   Bechara, A., Damasio, H., Tranel, D. & Damasio, A. R. Deciding advantageously before knowing the advantageous strategy. *Science.* 275, 1293–1295 (1997).

12.   Krain, A. L. *et al.* Distinct neural mechanisms of risk and ambiguity: a meta-analysis of decision-making. *Neuroimage.* 32, 477–484 (2006).

13. Bechara, A., Damasio, H., Damasio, A. R. & Lee, G. P. Different contributions of the human amygdala and ventromedial prefrontal cortex to decision-making. *J. Neurosci.* 19, 5473–5481 (1999).

14. Critchley, H. D., Mathias, C. J. & Dolan, R. J. Neural activity in the human brain relating to uncertainty and arousal during anticipation. *Neuron.* 29, 537–545 (2001).

15. Studer, B. & Clark, L. Place your bets: psychophysiological correlates of decision-making under risk. *Cogn. Affect. Behav. Neurosci.* 11, 144–158 (2011).

16. Crone, E. A., Somsen, R. J. M., Van Beek, B. & Van Der Molen, M. W. Heart rate and skin conductance analysis of antecendents and consequences of decision making.*Psychophysiology.* 41, 531–540 (2004).

17. Burke, C. J. & Tobler, P. N. Coding of reward probability and risk by single neurons in animals. *Frontiers in Neuroscience.* 5, 212 (2011).

18. Davies, A. C., Radford, A. N. & Nicol, C. J. Behavioural and physiological expression of arousal during decision-making in laying hens. *Physiol. Behav.* 123, 93–99 (2014).

19. Moe, R. O., Stubsjøen, S. M., Bohlin, J., Flø, A. & Bakken, M. Peripheral temperature drop in response to anticipation and consumption of a signalled palatable reward in laying hens (Gallus domesticus). *Physiol. Behav.* 106, 527–533 (2012).

20. Edgar, J. L., Lowe, J. C., Paul, E. S. & Nicol, C. J. Avian maternal response to chick distress. *Proc. Biol. Sci.* 278, 3129–3134 (2011).

21. Edgar, J. L., Nicol, C. J., Pugh, C. A. & Paul, E. S. Surface temperature changes in response to handling in domestic chickens. *Physiol. Behav.* 119, 195–200 (2013).

22. Moe, R. O. *et al.* Trace classical conditioning as an approach to the study of reward-related behaviour in laying hens: A methodological study. *Appl. Anim. Behav. Sci.* 121, 171–178(2009).

23. Zimmerman, P. H., Buijs, S. A. F., Bolhuis, J. E. &. Keeling, L. J. Behaviour of domestic fowl in anticipation of positive and negative stimuli. *Anim. Behav.* 81, 569–577 (2011).

24. Lowe, J. C., Abeyesinghe, S. M., Demmers, T. G. M., Wathes, C. M. & McKeegan, D. E. F. A novel telemetric logging system for

recording physiological signals in unrestrained animals. *Comput. Electron. Agric.* 57, 74–79 (2007).

25. Korte, S., Ruesink, W. & Blokhuis, H. Heart-rate variability during manual restraint in chicks from high- and low-feather pecking lines of laying hens. *Physiol. Behav.* 65, 649–652(1998).

26. Moberg, G. P. Biological response to stress: implications for animal welfare. In: *The Biology of Animal Stress.* (ed Moberg, G. P., Mench, J. A.), pp. 1–21. Wallingford: CAB International (2000).

27. Mendl, M., Burman, O. H. P. & Paul, E. S. An integrative and functional framework for the study of animal emotion and mood. *Proc. Biol. Sci.* 277, 2895–2904 (2010).

28. Bechara, A., Damasio, A. R., Damasio, H. & Anderson, S. W. Insensitivity to future consequences following damage to human prefrontal cortex. *Cognition.* 50, 7–15 (1994).

29. Davies, A. C., Nicol, C. J., Persson, M. E. & Radford, A. N. Behavioural and physiological effects of finely balanced decision-making in chickens. *PLOS ONE.* 9, e108809 (2014).

30. Browne, W. J., Caplen, G., Edgar, J., Wilson, L. R. & Nicol, C. J. Consistency, transitivity and inter-relationships between measures of choice in environmental

31. preference tests with chickens. *Behav. Processes.* 83, 72–78 (2010).

Chapter 4

Differentially Expressed Genes for Aggressive Pecking Behaviour in Laying Hens

Bart Buitenhuis, Jakob Hedegaard, Luc Janss, and Peter Sørensen

Aarhus University, Faculty of Agricultural Sciences, Department of Genetics and Biotechnology, Blichers allée 20, DK-8830 Tjele, Denmark

ABSTRACT

Background

Aggressive behaviour is an important aspect in the daily lives of animals living in groups. Aggressive animals have advantages, such as better access to food or territories, and they produce more offspring than low ranking animals. The social hierarchy in chickens is measured using the 'pecking order' concept, which counts the number of aggressive pecks given and received. To date, little is known about the underlying genetics of the 'pecking order'.

Results

A total of 60 hens from a high feather pecking selection line were divided into three groups: only receivers (R), only peckers (P) and mixed peckers and receivers (P&R). In comparing the R and P groups, we observed that there were 40 differentially expressed genes [false discovery rate (FDR)$P < 0.10$]. It was not fully clear how the 40 genes regulated aggressive behaviour; however, gene set analysis detected a number of GO identifiers, which were potentially involved in aggressive behavioural processes. These genes code for synaptosomes (GO:0019797), and proteins involved in the regulation of the excitatory postsynaptic membrane potential (GO:0060079), the regulation of the membrane potential (GO:0042391), and glutamate receptor binding (GO:0035254).

Conclusion

In conclusion, our study provides new insights into which genes are involved in aggressive behaviours in chickens. Pecking and receiving hens exhibited different gene expression profiles in their brains. Following confirmation, the identification of differentially expressed genes may elucidate how the pecking order forms in laying hens at a molecular level.

BACKGROUND

Aggressive behaviour in group-living animals is an important aspect of their daily lives, and this behaviour is partly used to establish social ranks in groups. Animals who rank highly in the social hierarchy have many advantages, such as better access to food and territories [1,2]. In studies in chickens, highly ranked males mated more often and produced more offspring than low ranking males [3]. Likewise, dominant hens produced more offspring over their lifespan than subordinate hens [4].

The social hierarchy in chickens can be measured by the number of aggressive pecks, which are usually aimed at the head of a receiving bird [5]. The onset of aggressive pecking differs between male and female chickens. Males initiate aggressive pecking behavior in their second week after hatching, and the pecking reaches adult levels when the chicken are eight to nine weeks old. Females initiate aggressive pecks at approximately five weeks of age, and they reach adult levels at nine to 10 weeks of age [6-8]. A stable hierarchy is established at approximately 20 weeks of age, and a number of different factors are involved in its formation. Kim and Zuk [9] demonstrated that previous social experience, parasite status, morphological characteristics and possibly age can be important factors in establishing a hen's rank in the group.

In the European Union, poultry are commonly housed in free range housing systems (Directive 1999/74/EC). Aggression can be a problem in these flocks and result in increased social stress. Additionally, skin damage can trigger cannibalism. The level of aggression has been shown to be lower in large groups of chickens than in small groups [10]. In order to reduce aggressive encounters under practical settings, it is important to identify the genes involved in aggressive pecking behaviour to understand how the pecking order is established in chickens.

To date, little is known about the underlying genetic mechanisms behind aggressive pecking in chickens. Previous selection experiments showed that aggressive pecking was not related to feather pecking because while the propensity to peck feathers changed during selection, there was no effect on the aggressive pecking behaviour (reviewed in [11]). There are indications that 'group selection' experiments for high and low production and survivability can influence aggressive behaviour in laying hens [12]. Later studies on these selection lines demonstrated that there were changes in the dopaminergic and serotonergic systems [13]. Animals injected with dopamine D2 receptor blockers showed a reduced frequency of aggressive pecks on subordinates [14]. Administration of 5-HT1-A and 5HT1-B antagonists resulted in increased aggressive pecks depending on the selection line [15]. Both the dopaminergic and serotonergic systems have been shown to influence aggressive behaviours in both mammals and birds [16-18].

The present study aimed to identify genes that regulate the aggressive pecking behaviour in chickens. In order to identify these genes, we compared the genome-wide profiles of chicken brain samples from aggressive and receiver hens using a 20 K chicken microarray. We tested the hypotheses that (1) differentially expressed (DE) genes are associated with the number of aggressive pecks given or received and (2) genes are DE among peckers, receivers and a mixed group of peckers and receivers.

RESULTS

Phenotype

The number of pecks given and pecks received per hen is shown in Figure 1. The number of pecks given during a three hour period ranged from 0 to 22, and the number of pecks received ranged from 0 to 46 (Figure 1). There was no difference between the cages in terms of the number of aggressive pecks performed per bird (Kruskal-Wallis $\chi^2_2 =$ 0.66, $P = 0.72$) or in the number of pecks received per bird (Kruskal-Wallis $\chi^2_2 = 1.34$, $P = 0.51$). Additionally, there was no difference in the animal weights between the cages ($F^1_{58} = 0.64$, $P = 0.44$).

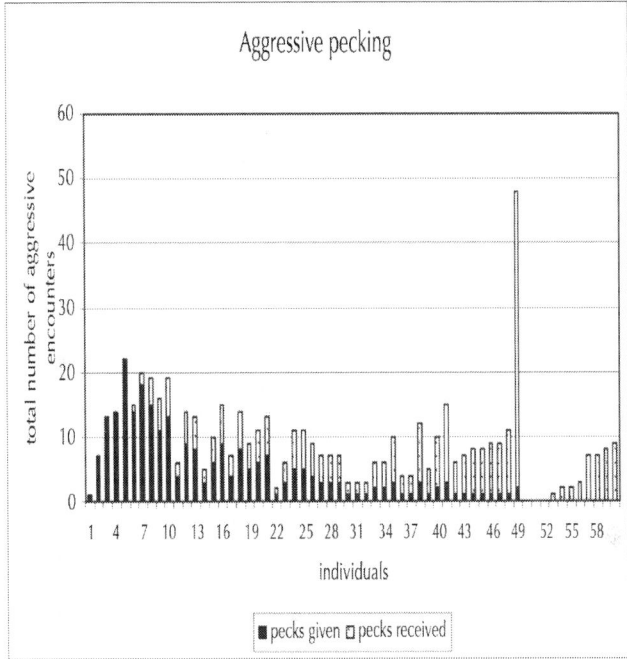

Figure 1: Histogram showing the number of aggressive pecks given and received by each hen during a 3 h social test.

Number of Pecks and Gene Expression

There was no relationship between the number of aggressive pecks given or received and the gene expression. The Spearman correlation ranged from -0.6 to 0.6, but the P value was between 0.99 and 1.

Grouping of Animals According to the Pecks Performed and Received

Group 1 consisted of 44 animals (peckers and receivers, P&R), Group 2 consisted of eight animals (receivers, R), Group 3 consisted of five animals (peckers, P), and Group 4 consisted of three animals. Group 4 was considered too small to perform a gene expression experiment on, so it was not used. There was no difference in body weight between the groups ($F^1_{58} = 0.39$, $P = 0.54$).

Gene Expression Analysis

In, the DE genes (n = 179) for the comparison between P&R and R are presented. The logFC was in the range of 1.5 to -2.5. Of the 179 genes, 106 had a gene annotation, and 148 were mapped to the chicken genome. However, none of these genes were significant at the false discovery rate (FDR) $P < 0.10$. In, the genes (n = 342) for the comparison between P&R and P are presented. The logFC was in the range of 1 to -4. Of the 342 genes, 218 had a gene annotation, 300 were mapped to the chicken genome, and 58 were significant at the FDR $P < 0.10$. In, the genes (n = 337) of interest from the comparison between R and P are presented. The logFC was in the range of 1.5 to -4. Of the 337 genes, 208 had a gene annotation, 301 were mapped to the chicken genome, and 40 were significant at the FDR $P < 0.10$. Figure 2 shows the logFC distribution of the significant genes for the P&R vs. P and R vs. P comparisons. Figure 3 shows a Venn diagram of the overlapping genes in the comparisons. There were 30 genes in common between the P&R vs. P and R vs. P comparisons.

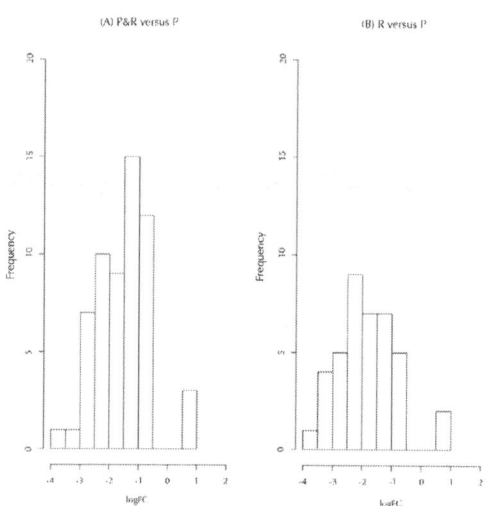

Figure 2: Histogram of the log fold changes (logFC) of the significantly (FDR $P < 0.1$) differentially expressed genes in group comparisons. Comparisons were made between (A) the pecker & receiver (P&R) and pecker (P) group, and between (B) the receiver (R) and the pecker (P) group.

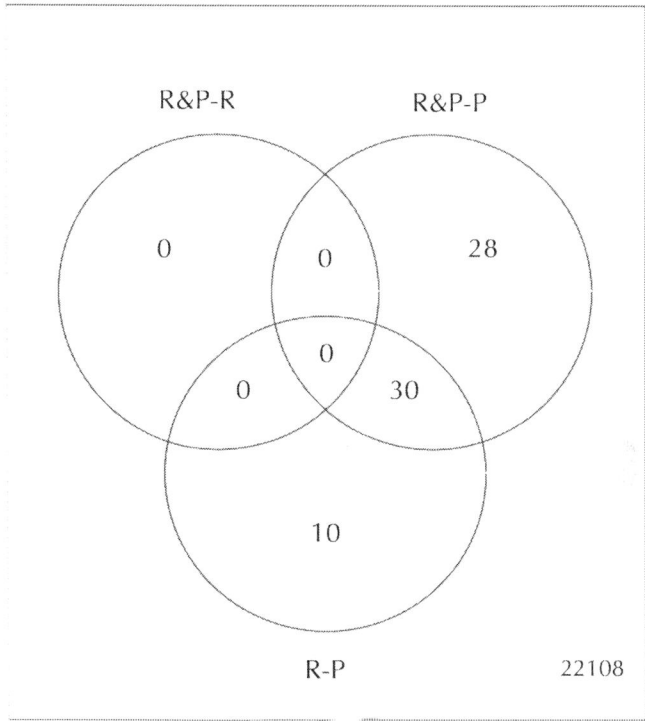

Overlap DE genes at adj.p.Value<0.1

Figure 3: Overlap of the significantly (FDR *P* < 0.1) differentially expressed genes between group comparisons. Venn diagram of the pecker & receiver (P&R) vs receiver (R) group, the P&R vs pecker (P) group, and the R vs P group

A heatmap of the 40 DE genes in the R vs. P comparison is shown in Figure 4. Clustering of the individuals based on their DE genes (FDR *P* < 0.10) showed that three out of the five peckers clustered together. The other two animals were assigned to two different clusters that contained receiver animals. Clustering of the individuals in the P&R vs. P comparison showed that the animals were mixed

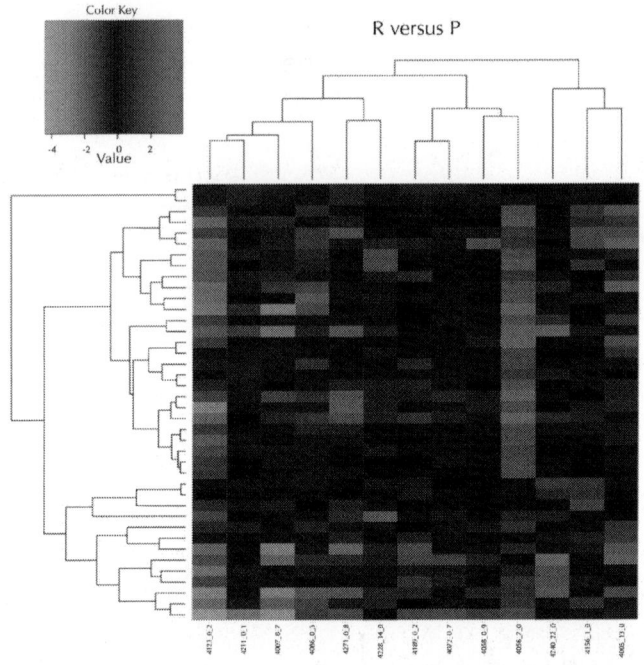

Figure 4: Heatmap of differentially expressed genes between the receiver (R) and pecker (P) group. There were 40 differentially expressed genes (FDR *P* < 0.1). The M values of the genes (rows) were ordered using the centred Pearson correlation and hierarchical clustering. The dendrogram shows the clustering results of the gene expression profiles. The arrays (columns) represent the individual hens, which are denoted with their id number, the number of aggressive pecks performed, and the number of aggressive pecks received (id_# pecks performed_# pecks received). The dendrogram shows the clustering results of hens based on the gene expression profiles. The red and green colours denote high and low intensities, respectively.

We tested for overrepresentation of gene sets representing biological processes (BP), cellular components (CC), and the molecular function (MF) of gene ontology (GO) in the P&R vs. P and R vs. P comparisons. For the R vs. P comparison, 33 GO identifiers were significant (P < 0.01). Of these 33 GO identifiers, 17 belonged to the BP set, 10 belonged to the CC set, and six belonged to the MF gene set (Figure 5). The GO identifiers that could potentially be involved in behavioural processes were related to synaptosomes (GO:0019797), the regulation of excitatory postsynaptic membrane potential (GO:0060079), the regulation of membrane potential (GO:0042391), and glutamate receptor binding (GO:0035254). The GO:0019797, GO:0060079, GO:0035254 genes were the glutamate receptors (*GRIN1, GRIN2A* and *GRIN2B*). The GO:0042391 genes were mainly acetylcholine receptors (*CHRNA1, CHRNA3, CHRNA4, CHRNB4*). GO identifiers that were involved in muscle development and lipid biosynthesis were also identified. The GO identifiers detected from the P&R vs. P comparison

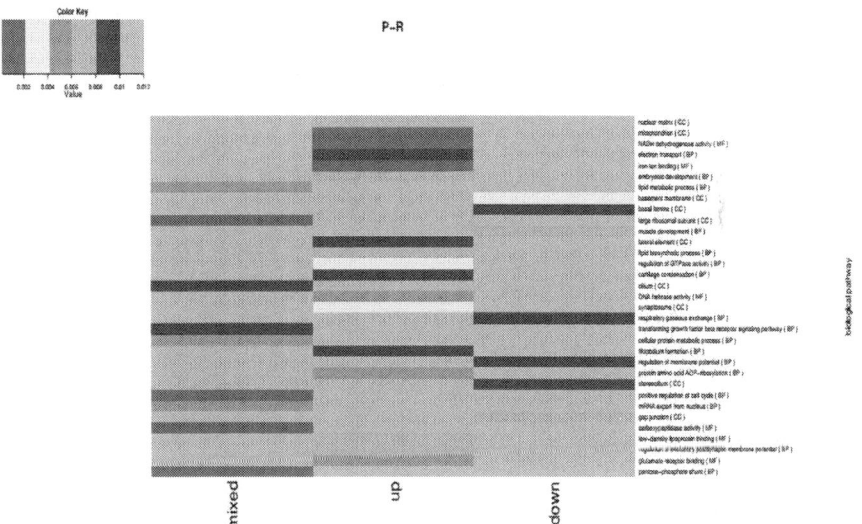

Figure 5: Representation of the significant GO identifiers detected in the comparison between the receiver (R) and pecker (P) group. The three alternatives (mixed, up, and down) are represented in the columns. The "mixed" alternative tested whether the genes in the set tended to be differentially expressed, without regard for the direction. In this case, the test is significant if the set mainly contains large test statistics, even if some results are positive and some are negative. The "up" alternative tested whether the genes in the set were up-regulated. The "down" alternative tested whether the genes in the set were down-regulated. The description of the biological pathways are listed in the rows and the GO class is listed in brackets (BP: biological processes, CC: cellular components, MF: molecular function). The colour gray denotes the GO identifiers *P* > 0.01.

DISCUSSION

This study identified genes that are involved in aggressive pecking behaviour in chickens, which is a behaviour related to social dominance. This research will help gain a better understanding of the underlying genetics of this behaviour. Our results showed that there was no association between number of pecks given or received and the gene expression level. However, comparison between the pecker and receiver animals showed that some genes were differentially expressed in the two groups.

Grouping of Animals

The aim of this study was to detect differences between pecker and receiver animals, and we assumed that the peckers were the aggressive animals and the receivers were the submissive animals or victims. In our study, we clearly divided groups of pecker (P) and receiver (R) animals. We are aware that there are animals in the pecker and receiver (P&R) group that have phenotypic profiles that may match either the pecker group or receiver group (Figure 1). However, it is difficult to assign a clear cut-off point to define an animal as a pecker or receiver when it both performs and receives aggressive pecks. Some researchers have used aggressive encounters (both pecks and receiving pecks) to rank animals in a social dominance hierarchy [19-22]. These approaches have varied, with some researchers counting only the number of pecks and receiving pecks [21], while others have taken the interactions (i.e. agonistic and avoidance behaviour) between animals into account [19,20,22]. Nevertheless, there are no major differences between the animal rankings in groups when the different ranking methods are used. In most cases, the animals defined as 'pure' peckers or receivers in our study were also among the top ranking or bottom ranking animals using the previously proposed formulas to rank animals in groups [19-22]. These similarities indicate that the groupings used in this study represent aggressive and submissive animals.

Genes and Behaviour

The P&R group had great variability in terms of the number of pecks given and received and is large compared to both the R and the P groups (Figure 1). This may explain why there were no significant genes at the FDR $P < 0.1$ level for the P&R vs. R comparison The P&R vs. P comparison detected 58 significant genes, and of these 30 overlapped with significant genes from the R vs. P comparison. This suggests that in both of these cases, the gene expression differences were caused by the P group.

Interpreting the DE genes in the R vs. P comparison is a first step towards understanding the underlying genetic mechanism behind aggressive behaviour in chickens. The R vs. P comparison identified 40 significant genes. It was not fully clear how the annotated genes were involved in regulating aggressive behaviour; however, as the annotation of the chicken genome improves, this may provide a clearer picture of how the genes are related to aggressive behaviour. None of the most obvious candidate genes for aggressive behaviour [see [16]] were identified in our comparisons. For example, it was previously shown that serotonergic receptors play a role in aggression in chickens [15]. It is possible that these genes were up- or down-regulated in a specific part of the brain, but could not be detected in our study of whole brain gene expression because the effect was diluted. Another explanation is that these genes influence aggressive behaviour via different allelic forms.

The gene set enrichment analysis, which involved all of the genes on the array, demonstrated that genes involved with muscle development were among the significant GO identifiers. In *Drosophila* tested for aggressive behaviour, genes involved in muscle contraction were among the significant GO identifiers [23]. It is not clear whether genes related to muscle development have the same function in the brain; however, in chickens, the largest bird usually ranks the highest [24]. In spite of this, we observed no difference in body weight between the pecking and receiver animals in our study.

Other GO identifiers detected in this study were involved in lipid metabolism, lipid synthesis and low density lipo-protein binding (Figure 5). Fatty acid binding proteins have been shown to play a role in the differentiation of neurons and glial cells in rats [25]. As a consequence of improved neural development, subsequent changes in the brain and behaviour could occur between low and high ranking animals, such as differences in the development of memory functions. In our study, some of the significant GO identifiers coded for synaptosomes, glutamate receptor binding, and the regulation of excitatory postsynaptic membrane potentials. Glutamate receptors, which were detected in the GO identifiers, play an important role in the development of memory formation following passive avoidance training in young chickens [26]. Therefore, memory may play a role in remembering the social hierarchy of the group.

CONCLUSIONS

In conclusion, our study provides a first insight into which genes are involved in aggressive behaviour in chickens. The results of our study showed that the level of expression is not dependent on the number of pecks given or received. It was not fully clear how the DE genes were involved in regulating aggressive behaviour. However, the gene set enrichment test showed that the DE genes coded for synaptosomes, and genes involved with lipid metabolism and memory formation. When confirmed in future studies, the DE genes may help scientists understand how the pecking order forms in laying hens at a molecular level.

METHODS

Chicken Lines, Tissue Sampling

In this study, we used 60 randomly selected laying hens from a high feather pecking selection line. This line was selected for eight generations for increased feather pecking behaviour based on a social feather pecking test, but the hens showed no difference in aggressive pecking behaviour when compared to the low feather pecking line [27]. The birds were reared in floor pens covered with a 5-cm thick layer of wood shavings. The temperature was 34°C when the hens were one day old, and it was gradually reduced to 20°C by the time the hens were eight weeks of age. The temperature was then maintained at 20°C for the remainder of the experiment. The light regime was 12 h light:12 h dark (12L:12D) from 0 to 14 weeks. Then one hour of light was added per week until the light regime was 16L:8D when the hens were 18 weeks of age. At 18 weeks, the chickens were transferred to four-bird battery cages in two levels. At 33 weeks of age, the chickens were randomly divided into three groups of 20 hens, and their body weights were measured. The birds had a week to adapt to the new environment and group composition. The groups consisted only of hens in order to replicate the commercial conditions that laying hens are maintained under. When the hens were 34 weeks of age, they were videotaped for 3 h between 14:00 h and 17:00 h. The 3 h time frame was chosen because this was the same time frame used during the selection procedure for the line. For each hen, the number of pecks given and received was scored. Aggressive pecks were defined as hard pecks aimed at the head or comb of the receiving bird, where no feather pulling was involved [5]. During the next day, between 8:00 h and 12:00 h, the 60 birds were decapitated. The whole brains were extracted and immediately frozen in liquid nitrogen and stored at -80°C for further use. The experiment has been performed according the regulation of the Danish Committee of Control with Animal Research (Dyreforsøgstilsynet).

Expression Profiling Using Microarrays

In total 60 samples from a high feather pecking line were used for the expression profiling experiments. The expression profiles of the 60 brains were measured using 20 K chicken oligo microarrays, which were printed and supplied by ARK-Genomics, Roslin Institute, UK via the European Animal Disease Genomics Network of Excellence (EADGENE) consortium. The arrays contained 20,678 oligos (64 to 70 mers), which corresponded to 20,640 chicken transcripts based on UMIST full length cDNA, DT40 full length cDNA, and ENSEMBL and TIGR ESTs Contigshttp://bioinformatics.roslin.ac.uk/eadgene/index. php/Chicken_-_Genomic_resources. More detailed descriptions of the 20 K chicken oligonucleotide microarrays are available at the National Center for Biotechnology Information's (NCBI's) Gene Expression Omnibus (GEO)[28,29], which is available through the GEO platform accession number GPL5480. Dual-channel microarray experiments were performed with a common reference design using total brain RNA that was purified from an unrelated animal as the reference. During the experiment, care was taken not to confound the factors of interest with the experimental batch sets. The whole brain was homogenized in liquid nitrogen using a TissueLyser (Qiagen-Retsch GmbH, 42781, Haan, Germany) fitted with 50 mL stainless steel grinding jars and 20 mm grinding balls. The total RNA was purified and treated with DNase treated using NucleoSpin RNA L (Macherey-Nagel GmbH & Co KG, 52355, Düren, Germany) following the enclosed protocol. The purified RNA samples were quantified using a NanoDrop ND-1000 spectrophotometer (NanoDrop Technologies, Thermo Fisher Scientific, Wilmington, DE 19810, USA), and the quality was evaluated by agarose gel electrophoresis. The samples were then stored at -80°C until use. From each sample, 10 μg of total RNA was labelled with Alexa-647, and 10 μg of the reference sample was labelled with Alexa-555 using the SuperScript Direct cDNA labelling System (Invitrogen, 2630, Taastrup, Denmark). The labelled cDNA was purified using the NucleoSpin 96 Extract II PCR Clean-up kit (Macherey-Nagel GmbH & Co KG, 52355, Düren, Germany). The labelled reference samples were mixed and divided into aliquots before combining the reference aliquots with the labelled samples. The slides were hybridised in six batches using a Discovery XT hybridization station (Ventana Discovery Systems, Tucson, AZ, USA) followed by scanning at a 5 μm resolution using the ScanArray Express HT system (version 3.0, Perkin Elmer, Waltham, MA, 02451, USA). Image analysis was conducted using GenePix Pro (version 6.0.1.27, Molecular Devices, Sunnyvale, CA, 94089-1136, USA) using irregular filled feature types and «MorphologicalClosingFollowedByOpening» background values. More detailed descriptions of the microarray experiment and data are available at the NCBI›s GEO[28,29] through the accession number GSE10380.

Statistical analysis of the microarray data was carried out in the R computing environment (version 2.5.0 for Windows) using the Linear Models for Microarray Analysis package (Limma, version 2.10.0, [30]), which is part of the Bioconductor project [31]. Spots flagged as «Not found» («Flags» = -50) by GenePix Pro and spots with either a «SNR 647» or «SNR 555» value less than 1 were excluded from the analysis by assigning the spot a weight of zero. The $\log_2$-transformed ratios of Alexa-647 to Alexa-555 (not background corrected median values) underwent within-slide normalizations using weighted loess with default parameters.

Statistical Analysis

The statistical analysis was performed in six steps. In Step 1, we tested for differences between cages in the level of aggressive pecking performed and received using the Kruskal-Wallis rank sum test (kruskal. test option in the R version 2.5.0). Differences in body weight between the cages and the rank order groups (see Step 2) were calculated using an analysis of variance (ANOVA) test.

In Step 2, we first examined the Spearman correlation between the expression level of each gene (M value) and the number of aggressive pecks given or received to test for equality to zero. The test was performed for 15,242 genes, and less than 30 values were missing. The P values were adjusted for multiple testing using the FDR procedure [32]. Second, the animals were grouped based on a combination of the number of aggressive pecks performed and received. The grouping of the animals may be performed in many different ways. From a practical point of view, we were interested in what made some birds peck (aggressive) and what made some birds be pecked (victim/ submissive). Therefore, the birds were assigned into groups based on a combination of the 'number of pecks performed' and the 'number of pecks received'. We assumed that the number of pecks performed was partly regulated by genes and that being a victim was regulated by different genes. This allowed the animals to be categorized into a group of 'pure' peckers and a group of 'pure' receivers (victims), which may show the difference between the peckers (P) and receivers (R). The group of birds which performed both pecking and receiving (P&R) was considered an intermediate. Group 1 contained animals that both pecked and received pecks. Group 2 contained animals that received pecks, but did not peck themselves. Group 3 contains animals that pecked, but did not receive any pecks. Lastly, Group 4 contained animals that did not peck and did not receive any pecks. Comparisons were made between Groups 1 and 2 (P&R vs. R), between Groups 1 and 3 (P&R vs. P), and between Groups 2 and 3 (R vs. P) using t-tests.

In Step 3, the differential expression of each gene was assessed using linear modelling and empirical Bayes methods, which were implemented using the R package Limma [30]. Test-pen was a fixed factor in the model. Each transcript targeted by a probe was tested for its expression change using a modified t-test. In the modified t-test, the residual standard deviations are moderated across the probe sets to ensure that there is a more stable inference for each transcript. The moderated standard deviations are a compromise between the individual transcript-wise standard deviations and the pooled standard deviation. Genes with an adjusted P value (FDR) < 0.1 are reviewed in the discussion, while the genes with P values < 0.01 are presented in 1, 2 &3.

In Step 4, the significant genes (FDR P < 0.10) were studied using a 2D-cluster analysis using the heatmap2 function from the gplots library (version 2.3.2) http://cran.r-project.org/web/packages/gplots/gplots.pdf The genes were clustered based on their normalized expression values using the correlation method.

In Step 5, the features on the arrays were annotated. For the annotations, we used 1) an annotation file available athttp://www.sigenae.org/fileadmin/_temp_/EADGENE_annotation/V2/EADGENE_oligo_annotation_GO_chicken_V2.csv*webcite*, and 2) a Unigene identifier, which was used to map the features on the array, and an annotation package, which was built using the Bioconductor package AnnBuilder (version 1.14.0).

In Step 6, each gene on the array was assigned to a GO identifier, and a gene set enrichment test was performed to compare P&R vs. P and R vs. P using Limma [30,33]. This is a modified version of the gene set enrichment test reported by Mootha et al [34]. For this test, it is not necessary to make a hard cut-off point between the genes that are DE and those that are not [34]. The method is especially useful in for traits that are influenced by many genes, which each have a small effect, like behavioural traits.

AUTHORS' CONTRIBUTIONS

Conceived and designed the experiment: BB. Performed the microarray experiment: JH Analyzed the data: BB, PS, and LJ Wrote the paper: BB. All authors contributed to the discussion of the results and agreed on the contents of the paper.

ACKNOWLEDGEMENTS

We kindly acknowledge Richard Talbot and his colleagues (ARK-Genomics, Roslin Institute, Edinburgh, UK) for providing the arrays used in this study. We thank Birte Nielsen (Aarhus University, DK) and Bas Rodenburg (Wageningen University, NL) for reading an earlier version of this manuscript, Kirsten Lund Balthzersen for taking care of the animals, Jørgen Kjær, Poul Sørensen, and Bodil Hjarvard for their help in obtaining the brain samples, and Mette Lindstrøm Bech for her assistance in making the behavioural observations. This study was supported by a grant from the Danish Ministry of Science, Technology and Innovation (FTP no. 274-05-0239).

REFERENCES

1.   Hall CL, Fegigan LM: Spatial benefits afforded by high rank in white-faced capuchins. *Anim Behav* 1997, 53:1069-1082.

2.   Lahti K, Koivula K, Rytkönen S, Mustonen T, Welling P, Pravosudov VV, Orell M: Social influences on food caching in willow tits: a field experiment. *Behav Ecol* 1998, 9:122-129.

3.   Jones ME, Mensch JA: Behavioral correlates of male mating success in a multi-sire flock as determined by DNA fingerprinting. *Poult Sci* 1991, 70:1493-1498.

4.   Collias N, Collias E, Jennrich RI: Dominant red junglefowl (*Gallus gallus*) hens in an unconfined flock rear the most young over their lifetime. *Auk* 1994, 111:869-872

5.   Savory CJ: Feather pecking and cannibalism. *World Poult Sci J* 1995, 51:215-219.

6.   Guhl AM: The development of social organization in the domestic chick. *Anim Behav* 1958, 6:92-111.

7.   Kruijt JP: Ontogeny of social behaviour in Burmese red junglefowl (*Gallus gallus spadiceus*). *Behaviour* 1964, 12:1-201.

8.   Rushen J: How peck orders of chickens are measured: a critical review. *Appl Anim Ethol* 1984, 11:255-264.

9.   Kim T, Zuk M: The effect of age and previous experience on social rank in female red jungle fowl (*Gallus gallus spadiceus*). *Anim Behav* 2000, 60:239-244.

10. Rodenburg TB, Koene P: The impact of group size on damaging behaviours, aggression, fear and stress in farm animals. *Appl Anim Behav Sci* 2007, 103:205-214.

11. Buitenhuis AJ, Kjaer JB: Long term selection for reduced or increased pecking behaviour in laying hens. *World Poult Sci J* 2008, 64:477-487.

12. Muir WM: Group selection for adaptation to multiple-hen cages: selection program and direct responses. *Poult Sci* 1996, 75:447-458.

13. Cheng H, Muir WM: The effect of genetic selection for survivability and productivity on chicken physiological homeostasis. [http://journals.cambridge.org/ action/displayFulltext?type=6&fid=619 164&jid=WPS&volumeId=61&issueId=03&aid=619160&fulltex tType=RV&fileId=S00439 33905000279] *World Poult Sci J* 2005, 61:383-397

14. Dennis RL, Muir WM, Cheng HW: Effects of raclopride on aggression and stress in diversely selected chicken lines. *Behav Brain Res* 2006, 175:104-111.

15. Dennis RL, Chen ZQ, Cheng HW: Serotonergic mediation of aggression in high and low aggressive chicken strains. *Poult Sci* 2008, 87:612-620.

16. Nelson RJ, Chiavegatto S: Molecular basis of aggression. *Trends Neurosci* 2001, 24:713-719

17. Van Hierden YM, Korte SM, Ruesink EW, Van Reenen CG, Engel B, Korte-Bouws GAH, Koolhaas JM, Blokhuis HJ: Adrenocortical reactivity and central serotonin and dopamine turnover in young chicks from a high and low feather-pecking line of laying hens. *Physiol and Behav* 2002, 75:653-659.

18. Kjaer JB, Hjarvard BM, Jensen KH, Hansen-Møller J, Naesby-Larsen O: Effects of haloperidol, a dopamine D2 receptor antagonist, on feather pecking behaviour in laying hens. *Appl Anim Behav Sci* 2004, 86:77-91.

19. Lee YP, Craig JV, Dayton AD: The social rank index as a measure of social status and its association with egg production in White Leghorn pullets. *Appl Anim Ethol* 1982, 8:377-390.

20. Mendl M, Zanella AJ, Broom DM: Physiological and reproductive correlates of behavioural strategies in female domestic pigs. *Anim*

Behav 1992, 44:1107-1121.

21. Galindo F, Broom DM: The relationship between social behaviour of dairy cows and the occurrence of lameness in three herds. *Res Vet Sci* 2000, 69:75-79.

22. Lamprecht J: Social dominance and reproductive success in a goose flock (*Anser indicus*). *Behaviour* 1986, 97:50-65.

23. Edwards AC, Rollmann SM, Morgan TJ, Mackay TFC: Quantitative genomics of aggressive behaviour in *Drosophila melanogaster*. *PLoS Genetics* 2006, 2:e154.

24. Pagel M, Dawkins MS: Peck orders and group size in laying hens: 'futures contrasts' for non-aggression. *Behav Processes* 1997, 40:13-25.

25. Bennett E, Stenvers KL, Lund PK, Popko B: Cloning and characterization of a cDNA encoding a novel fatty acid binding protein from rat brain. *J of Neurochem* 1994, 63:1616-1624.

26. Salenska EJ, Chaudhury D, Bourne RC, Rose SPR: Passive avoidance training results in increased responsiveness of voltage and ligand-gated calcium channels in chick brain synaptpneurosomes. *Neuroscience* 1999, 93:1507-1514.

27. Kjaer JB, Sørensen P, Su G: Divergent selection on feather pecking behaviour in laying hens (*Gallus gallus domesticus*). *Appl Anim Behav Sci* 2001, 71:229-239

28. Edgar R, Domrachev M, Lash AE: Gene Expression Omnibus: NCBI gene expression and hybridization array data repository. *Nucl Acids Res* 2002, 30:207-210

29. Barrett T, Suzek TO, Troup DB, Wilhite SE, Ngau WC, Ledoux P, Rudnev D, Lash AE, Fujibuchi W, Edgar R: NCBI GEO: mining millions of expression profiles--database and tools. *Nucl Acids Res* 2005, (33 Database):D562-D566.

30. Smyth GK: Linear models and empirical bayes methods for assessing differential expression in microarray experiments. *Stat Appl Genet Mol Biol* 2004., 3)

31. Gentleman RC, Carey VJ, Bates DM, Bolstad B, Dettling M, Dudoit S, Ellis B, Gautier , Ge Y, Gentry J, Hornik K, Hothorn T, Huber W, Iacus S, Irizarry R, Leisch F, Li C, Maechler M, Rossini AJ, Sawitzki G, Smith C, Smyth G, Tierney L, Yang JY, Zhang J: Bioconductor: open software development for computational

biology and bioinformatics. *Genome Biol* 2004, 5:R80.

32.   Benjamini Y, Hochberg Y: Controlling the false discovery rate. *J R Stat Soc B* 1995, 57:289-300.

33.   Michaud J, Simpson KM, Escher R, Buchet-Poyau K, Beissbarth T, Carmichael C, Ritchie ME, Schütz F, Cannon P, Liu M, Shen X, Ito Y, Raskind WH, Horwitz MS, Osato M, Turner DR, Speed TP, Kavallaris M, Smyth GK, Scott HS: Integration of RUNXI downstream pathways and target genes. *BMC Genomics* 2008, 9:363

34.   Mootha VK, Lindgren CM, Eriksson K-F, Subramanian A, Sihag S, Lehar J, Puigserver P, Carlsson E, Ridderstråle M, Laurila E, Houstis N, Daly MJ, Patterson N, Mesirov JP, Golub TR, Tamayo P, Spiegelman B, Lander ES, Hirschhorn JN, Altshuler D, Groop LC: PGC-1α-responsive genes involved in oxidative phosphorylation are coordinately downregulated in human diabetes. *Nature Genet* 2003, 34:267-273.

Early-Light Embryonic Stimulation Suggests a Second Route, Via Gene Activation, to Cerebral Lateralization in Vertebrates

Cinzia Chiandetti[1], Jessica Galliussi[1, 2], Richard J. Andrew[3], and Giorgio Vallortigara[1]

[1]CIMeC – Center for Mind/Brain Sciences. University of Trento

[2]Department of Life Science - Psychology Unit "Gaetano Kanizsa". University of Trieste

[3]Life Sciences, University of Sussex

ABSTRACT

Genetic factors determine the asymmetrical position of vertebrate embryos allowing asymmetric environmental stimulation to shape cerebral lateralization. In birds, late-light stimulation, just before hatching, on the right optic nerve triggers anatomical and functional cerebral asymmetries. However, some brain asymmetries develop in absence of embryonic light stimulation. Furthermore, early-light action affects lateralization in the transparent zebrafish embryos before

their visual system is functional. Here we investigated whether another pathway intervenes in establishing brain specialization. We exposed chicks' embryos to light before their visual system was formed. We observed that such early stimulation modulates cerebral lateralization in a comparable vein of late-light stimulation on active retinal cells. Our results show that, in a higher vertebrate brain, a second route, likely affecting the genetic expression of photosensitive regions, acts before the development of a functional visual system. More than one sensitive period seems thus available to light stimulation to trigger brain lateralization.

INTRODUCTION

Asymmetry along the left-right axis is a feature common to all vertebrates. Heart and liver are placed to the left and right side respectively[1, 2, 3, 4] and even paired-symmetric organs display some degree of asymmetry (e.g. lungs differ in the number of lobes). The brain exhibits profound anatomical and functional asymmetries (review[5]). How anatomical asymmetry is imposed on a seemingly bilaterally symmetric structure, the vertebrate neural tube, is however still largely obscure. Selective expression of the transforming growth factor (TGF) family member Nodal, a signal transduction pathway, on the left side of the early embryo seems to mediate the asymmetrical morphogenesis and placement of the internal organs through activation of a signaling cascade[6, 7]. Whilst such a Nodal cascade controls ventral forebrain development, its effect on lateralization is on epiphyseal gene expression in the dorsal forebrain (e.g. in zebrafish, the asymmetry of the diencephalic habenular nuclei and the photoreceptive pineal complex[8, 9]).

Lateralization mediated by Nodal cascade seems to operate in the brain, triggered by asymmetric sensory stimulation in embryo. The processes underlying the asymmetric morphology and positioning of the viscera are accompanied by a slight torsion of the embryo whose forehead points to the right[10]. Such a rightward spinal torsion seems to occur in all amniotes[11], including human embryos, which also display a right-turn of their head[12]. Asymmetric turning associated with Nodal signals may set the stage for either direct asymmetrical sensory stimulation of the embryo (because of its placement in utero or in ovo[13, 14]) or by constrained motor patterns that in turn may promote

asymmetrical stimulation (for instance, a slight preference to move the right arm because of turning of the embryo can then be enhanced by an increased eye-hand contact on the right side[15]).

This has been investigated in detail in the avian brain. During incubation, the birds' embryos bend so that the head is asymmetrically tilted with the right eye placed below the egg surface and the left eye leant below the wing[16]. In this position, environmental light penetrating the eggshell acts on the retina of the right eye only, producing structural asymmetries on the ascending visual projections and thus modulating functional cerebral specialization[14, 17, 18, 19]. Environmental illumination during latest stages of embryonic development is crucial in the determination of brain asymmetries through the selective action on the fully-formed visual receptors[20]: domestic chicks hatched from dark incubated eggs lack any asymmetry in the visual pathways[21, 22]. Furthermore, swapping the exposed eye by withdrawing the embryo's head from the egg (i.e., making the left rather than the right eye to receive illumination) reverses the pattern of asymmetry in both chicks[23, 21] and pigeons[24, 25]. Environmental illumination affects also functions of the left (unstimulated) eye and associated contralateral brain structures (as shown in attack, copulation and detection of predator[23] and visuo-spatial abilities[26]) by modifying inter-hemispheric cross-talk[27, 28].

Despite light being such a strong environmental trigging factor, some asymmetries in birds are unaffected by embryonic light exposure. For instance, uni-hemispheric sleep patterns[29], lateralized mechanisms of social recognition[30], and the neural mechanisms underlying imprinting[31, 32] are apparent also in dark-incubated birds. Besides, work on zebrafish has shown only partial correspondence between the reversal of the visceral situs and diencephalic asymmetries and the reversal of lateralized behaviours[33]. Similarly, in rare cases of spontaneous situs inversion in humans only some of the brain and behavioural lateralities change the direction of their asymmetry (e.g. language dominance continued to be a feature of the left hemisphere, i.e., not reversed[34, 35]). All this is suggestive of multiple genetic/environmental routes to brain lateralization.

In zebrafish environmental illumination applied early in development (at day one after fertilization) is needed to generate left/right cerebral asymmetries. After light stimulation, the left eye shows more interest in motivating stimuli[36, 37, 38]. At this early stage of development, light

reaching the embryos is not acting over photoreceptive cells in the retina, because these have not yet differentiated[39]. Rather, since larval zebrafish are transparent, light may influence the genetic expression of undifferentiated cells of photosensitive regions[40].

The evidence discussed insofar is suggestive of at least two different developmental pathways for determination of laterality in the vertebrate brain. The first pathway may involve genes of the Nodal cascade which determine a rightward torsion of the embryo that allows asymmetric light stimulation to trigger anatomical and functional asymmetries. A second pathway, never demonstrated in higher vertebrates (birds and mammals), may act directly by light stimulating the embryo at an age before the development of a functional visual system. The effects of such a second pathway, if proved, may explain why some forms of laterality seem to be unaffected by embryonic late stimulation.

Here we show for the first time that early environmental illumination provided to chicks' embryo at an age in which light cannot exert any effect on fully-formed retinal receptors which seem not to be in place[41] may nonetheless cause cerebral lateralization, likely involving the second developmental pathway, i.e., by affecting the genetic expression of undifferentiated cells of photosensitive regions.

RESULTS

In Experiment 1, chicks incubated in darkness (Di-chicks), exposed to light during the first 3 days after fertilization (EarlyLi-chicks) and the last 3 days before hatching (LateLi-chicks) were free to peck at food grains, scattered in an array of identical vertical sectors: a central one, 8 left and 8 right sectors (Figure 1). The total amount of pecks in each sector over a 3 minutes period was scored for each chick. In a repeated measures ANOVA, Hatch (Di-, EarlyLi- and LateLi-chicks) Side (Left and Right) and Distance (1 to 8 sectors) were analyzed as factors. The ANOVA showed a significant effect of Hatch ($F_{(2,85)}$ = 4.046, P = 0.021) and a significant effect of Distance ($F_{(7,595)}$ = 1093.597, P < 0.001) with decreased pecking with increasing distance from the center; there was also a significant main effect of Side ($F_{(1,85)}$ = 10.657, P = 0.002). The interaction between Side and Hatch was significant ($F_{(2,85)}$ = 4.841, P = 0.010) with Di-chicks choosing equally for the left and the right side of the grid (Di- toward left: 2.353 ± 0.128; toward right: 2.427 ± 0.156,

$t_{(28)}$ = 0.533, P = 0.598, Paired Samples t-Test) and both EarlyLi- and LateLi-chicks preferring significantly the left side (respectively, EarlyLi-: toward left: 2.531 ± 0.133; toward right: 2.165 ± 0.133, $t_{(27)}$ = −3.731, P = 0.001; LateLi-: toward left: 2.907 ± 0.067; toward right: 2.573 ± 0.100, $t_{(30)}$ = −3.599, P= 0.001) as shown in Figure 2. The interaction between Distance and Hatch was significant ($F_{(14,595)}$ = 3.322, $P <$ 0.001), as well as the interaction between Side and Distance ($F_{(7,595)}$ = 4.173, $P < 0.001$).

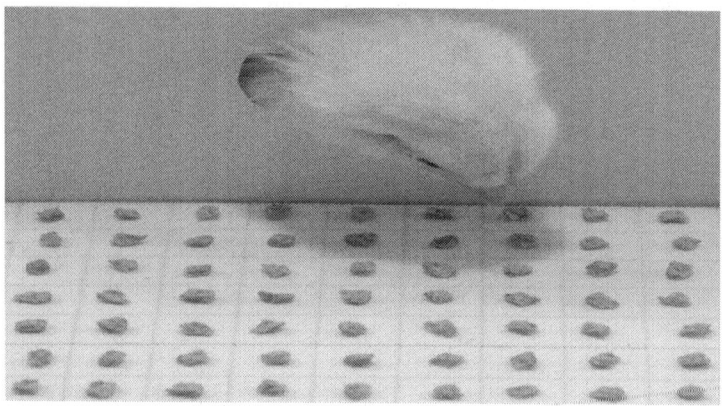

Figure 1: The chick's head and neck protruding from the window of the confining box and oriented toward its left during the activity of pecking at the cancellation grid.

Single grains of food are homogeneously disposed every cm on a double sided sticky tape.

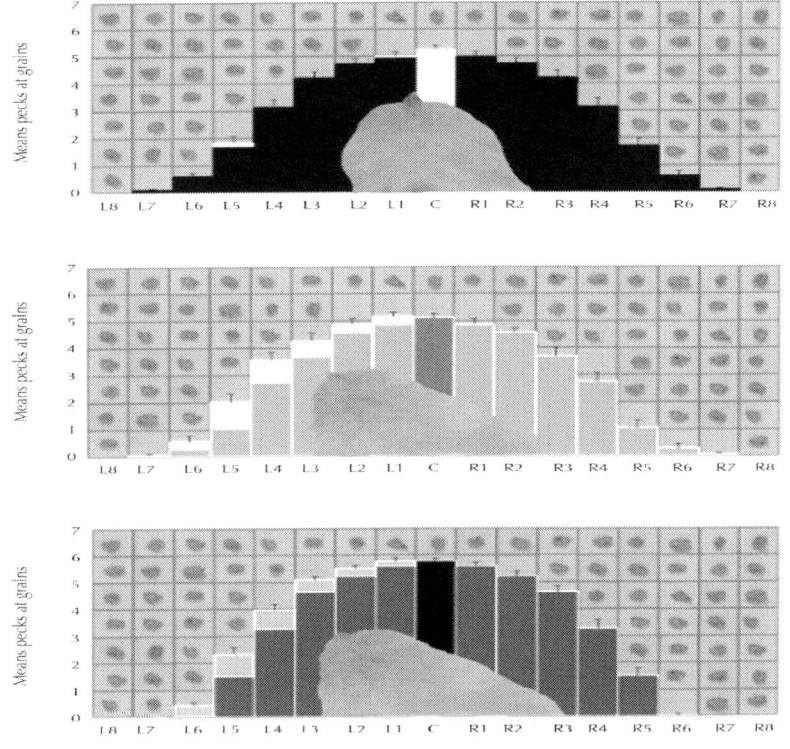

Figure 2: Means (+ s.e.m.) pecks at grains made by Di-chicks (top), EarlyLi-chicks (middle) and LateLi-chicks (bottom) at the cancellation grid toward left (L) and right (R) from the central midline (C) in 3 minutes of activity.

Lighter parts indicate the greater amount of pecks toward left. Single grains of food and the chick's head viewed from above are superimposed in post-editing to the graph for representational purposes only.

In Experiment 2, Di-, EarlyLi- and LateLi-chicks were left free to run from a starting box to a feeder located on the opposite ends of a long runway (Figure 3(a)). The routes covered by the chicks were scored with a video analysis software (Videopoint®) and then analyzed with Matlab® to determine the direction (left *vs*. right) of each route.

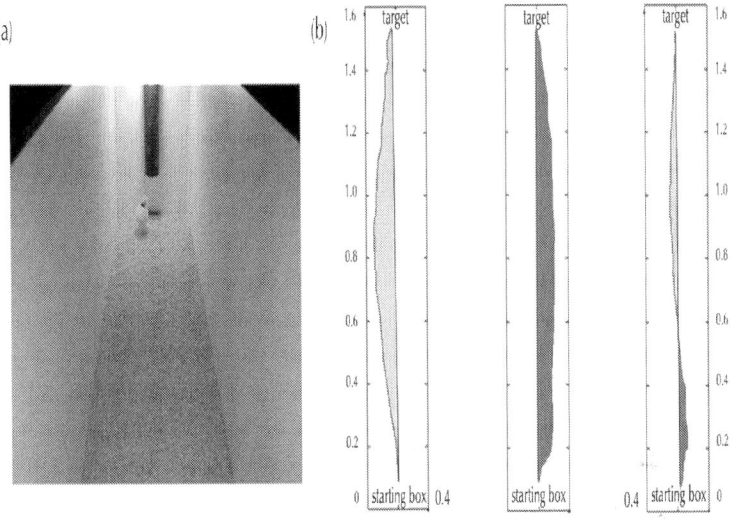

Figure 3: A chick arriving at the target feeder located below a conspicuous landmark (a), three examples of routes (in red) from the starting box to the target as analyzed to establish the route direction (b).

The black line connecting the ends of the route represents the optimal route. Length and width of the apparatus are expressed in metre.

We estimated the area (expressed in m^2) between the real route covered by the chick and the optimal route both for leftward and rightward trajectories as visible in Figure 3(b). A repeated measures ANOVA with Hatch (Di-, EarlyLi- and LateLi-chicks) and Route Direction (Left *vs.* Right) as factors showed no difference across hatching conditions ($F_{(2,58)}$ = 1.115, P = 0.335). By contrast, the main factor Route Direction was significant ($F_{(1,58)}$ = 153.913, $P < 0.001$) with all animals running more to the left than to the right (respectively, 0.049 m^2 ± 0.006 *vs.* 0.016 m^2 ± 0.004; $t_{(60)}$ = 12.514, $P < 0.001$, Paired Samples t-Test). When an obstacle was inserted in the center of the runway (along the midline of the optimal straight route as shown in Figure 4(a)) and the chick was allowed to reach the target 8 consecutive times, an ANOVA with Hatch as between-subject factor and Detour Direction as dependent variable revealed a significant heterogeneity between groups ($F_{(2,54)}$ = 3.626, P = 0.033). Di-chicks chose to detour the obstacle leftward ($t_{(19)}$ = 3.644, P = 0.002, One-Sample t-Test), whereas EarlyLi- and LateLi-chicks showed no systematic preference for a direction (respectively, $t_{(17)} = -0.676$, $P = 0.508$; $t_{(18)} = 0.468$, $P = 0.645$) as shown in Figure 4(b).

(a)   (b)

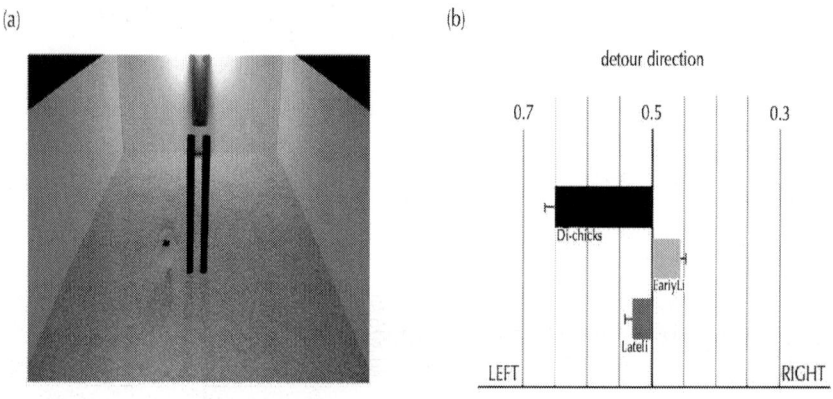

Figure 4: A chick on the left side of the obstacle while reaching the target (a), Di-chicks choose to detour the obstacle on the left significantly more often than EarlyLi- and LateLi-chicks (b).

DISCUSSION

Chicks hatched from eggs exposed to ambient illumination for three days either at an early or a late stage of embryonic development performed in a comparable way in two different tasks. In the first experiment, both EarlyLi- and LateLi-chicks showed a leftward bias when allowed to peck at crumbles scattered on a grid in front of them. Chicks maintained in dark, in contrast, directed their pecks uniformly toward the left and the right hemi-spaces. The result that LateLi-chicks have a left bias in attending the target position whereas Di-chicks show no systematic asymmetrical preference confirms previous results[25]. Such a late light action is well-known[14] and it is thought to affect brain structures by a retinal route, stimulating photoreceptors of the right eye during a critical period (last three days before hatching) in which retinal ganglion cells start to be active.

The result that an early light stimulation may induce a comparable behavioural lateralization provides the first evidence that illumination also play a role on structures of the nervous system that are not the fully-formed eye.

In the second experiment, we found again a striking similarity of the effects of the early and the late exposure to light. Both EarlyLi- and LateLi-chicks showed no bias when they had to detour an obstacle in order to reach a target, moving around either to the left or the right equally. On the contrary, Di-chicks showed a bias, choosing to detour the obstacle significantly more often to the left. Note that both chicks stimulated by light and chicks incubated in the dark run in a comparable way when they had only to reach for a target, meaning that there were no differences in motor behaviour depending on the different incubation conditions. Rather, a difference in control of attention may explain the results. Both EarlyLi- and LateLi-chicks, having decided on approach to the target, are able to ignore the obstacle altogether, whereas Di-chicks have to actively sustain approach to the target by using the right eye to view the obstacle. Di-chicks seem less able in sustaining attention than the groups of chicks exposed to light much as light chicks use the right eye in initial selection of target. There is evidence that many vertebrates view potential danger with the right eye in order to sustain examination and assessment (reviews[5, 42]).

Here we showed that two different lateralized behaviours are affected by environmental illumination: both the early and the late stimulation seem to affect the lateralized behaviours in the same direction. Despite the observed effects involve visuo-spatial behaviours mediated primarily by visual processing, functions like sustained attention or inhibition of responses are also crucial in performing these tasks. These functions may be differently modulated by asymmetries of other brain regions than the visual pathways on which the late light stimulation operates. The responsible mechanism of the early stimulation needs to be investigated, but we hypothesize that one route to the development of cerebral asymmetries may involve the genetic expression of photosensitive regions.

The early photosensitivity in the epiphyseal area of zebrafish has been proposed as responsible for this early action of light, via gene activation, in cells which are not specialized retinal photoreceptors[38, 39]. It has been demonstrated that these photosensitive cells respond to light very early in fish embryonic development[43], well before retinal photoreceptors[44, 45]. To our knowledge it is not known whether the epiphyseal photoreceptors are active during the first 3 days of embryonic development in the chick, though it can be reasonably assumed that a similar developmental pattern may be shared across vertebrates. However, it is not possible to exclude that other photosensitive molecules in the developing telencephalon and diencephalon in early stages of development may be directly involved[46].

We know that during the time-window in which we applied the early-light, the chick embryo still lies in a symmetric position within the egg[41] but asymmetric processes are likely to be already at play. Kuan and collaborators[38] report that lateralized Nodal signaling influences the directional asymmetry of the parapineal which in turn mediates habenular asymmetry in fish. The asymmetry of early action of light may be due to asymmetry of gene expression, presumably in the area where the parapineal develops and where photoreceptors start to be active early. A comparable cascade may be at work in the chick brain, possibly involving different areas outside the habenula. Indeed, in this species the epiphysis is forming as a separate knob at 60 hours after fertilization[41]. The epiphyseal area could be the target substrate for the early illumination to trigger visuo-spatial asymmetry comparable to the late activity of light on the optic nerve.

Although this explanation has to be taken cautiously, it is clear that structures other than the eye may be modulated by light exposure. Evolution may have provided organisms with more than one sensitive temporal window as a chance for the light to exert its important ecological role in triggering brain asymmetry.

METHODS

The study was carried out in compliance with the European Community and the Italian law on animal experiments by the Ministry of Health, under the authorization of the Ethical Committee of the University of Trieste (protocol number 385 pos II/9 dd 16.03.2012).

Subjects

Two hundred and six Hybro (White Leghorn) chicks (*Gallus gallus*) hatched in our laboratory under controlled conditions were used. The eggs were obtained from a local commercial hatchery immediately after fertilization; thereafter, some eggs (n = 70) were kept in complete darkness until the hatching day (Di-chicks) in an incubator FIEM snc, MG 100 H (45 cm wide × 58 cm high × 43 cm deep), under controlled temperature (36.7°C) and humidity (about 50–60%) conditions; some eggs (n = 66) were exposed to light from day 1 to day 3 (nearly 70 hours) after fertilization (EarlyLi-chicks) and thereafter remained in the dark; some others (n = 70) were maintained in darkness and exposed to light from day 18 of incubation (LateLi-chicks) to day 21 of hatching. A 60 W incandescent light bulb provided light within of the incubator. Immediately after hatching, chicks were reared singly in metal cages (22 cm wide × 30 cm high × 40 cm deep) illuminated by 30 W fluorescent lights (12 L: 12 D cycle) and located in a separate room at 28–30°C. Food and water were available *ad libitum*. Chicks of each incubation condition were tested on day 4 post-hatch. The experimenter was blind to the hatching condition.

Cancellation Task

Apparatus. The apparatus was the same used in previous experiments[47, 26] and consisted of a white uniform wooden enclosure (50 cm wide × 45 cm high × 50 cm deep) with a brown ground. A white cardboard box (14 cm wide × 14 cm high × 14 cm deep) served as confining box; it was fixed at the middle of the rear wall and presented on its frontal wall a circular window measuring 2 cm in diameter. A Poliplack® array positioned centrally and exactly beyond the window in the cardboard box, 4 cm above the floor, was divided in 170 compartments of 1 × 1 cm (17 columns of 10 compartments each) containing a single grain of chicks crumbles each. It was covered by a double sided sticky tape which provided grains to remain in a fixed position while the chick was pecking at close elements. A lamp of 30 W placed exactly on the top of the cardboard box illuminated the apparatus. The behaviour was videorecorded by a Panasonic® NV-GS27 camera connected to a monitor so that animals› activity could be observed without interference.

Procedure

After 3 h of food deprivation, on day 3 of age, each chick was in turn placed within the confining box and accustomed to protrude head and neck through the circular window in order to feed from a rectangular dish located frontally outside. The chick is motivated to do this because the environment within the box is dark while the external surroundings well lit; the chick spontaneously comes out and goes back inside the confining box. After 15 minutes of activity the chick was brought back in its rearing cage with food and water available until the evening. The next day, after overnight of food deprivation only, the chick was placed in the confining box and observed for a total of 3 minutes during which time it was free to peck at the grains of food regularly scattered in the above described array. The body-restrained condition ensured a continuous alignment with the midline of the searching area used in the test.

Running task

Apparatus. The apparatus consisted of a white uniform rectangular wooden enclosure (40 cm wide × 50 cm high × 160 cm deep) with sawdust (5 cm in depth) on the floor.

At the middle of the opposite smaller ends of the enclosure were fixed a white Poliplack® starting box (12 cm wide × 12 cm high × 12 cm deep) with no frontal wall and a landmark (a blue and red cardboard cylinder) placed 7 cm above the floor. The landmark indicated the presence of a small yellow and green rectangular plastic feeder (target) exactly below it. A uniform illumination of the whole apparatus was provided by two lamps of 25 W placed exactly on the top of the starting box and of the landmark. The behaviour was videorecorded by a Panasonic® NV-GS27 camera and scored offline. In order to keep track of the chick›s movements within the apparatus, a black removable piece of paper was temporarily attached on the chick›s back.

Procedure

On day 3 of age, after 3 hours of food deprivation, all the chicks were first positioned inside the starting box and left free to move around in order to get acquainted with the novel environment by reaching the target and finding mealworm larvae (*Tenebrio molitor*). Chicks did this quite spontaneously since the landmark was conspicuous and caught their interest. The habituation phase lasted in average 30 minutes: each chick was placed in the starting box and left free to walk toward the landmark to find the reward. The next day, after overnight of food deprivation, each chick was in turn placed within the apparatus in the starting box and left free to run toward the landmark once. No food was available in the feeder. The trial started as soon as the chick came out of the starting box and ended when the chick rested at the feeder. The positions of the starting box and the target were counterbalanced between subjects.

The same procedure was used for the animals that underwent the running task in presence of the obstacle placed in the centre of the apparatus (this time 80 cm wide). The obstacle consisted of 2 identical black plastic cylinders (2 cm in diameter and 30 cm high), spaced 2 cm from one another and blocked by a plastic bar positioned just underneath the sawdust. The proportion of the left routes over the total 8 routes was calculated to determine detour direction.

AUTHOR'S CONTRIBUTIONS

C.C. and G.V. wrote the manuscript text. C.C., G.V. and R.J.A. conceived the experiments. C.C. and J.G. performed the experiments, analyzed the data and prepared the figures. C.C., R.J.A. and G.V. reviewed the manuscript.

ACKNOWLEDGEMENTS

This work has been realized thanks to the support from the Provincia Autonoma di Trento and the Fondazione Cassa di Risparmio di Trento e Rovereto. We are grateful to Tommaso Mastropasqua and Tommaso Pecchia for technical assistance.

REFERENCES

1. Burdine, R. D. & Schier, A. F. Conserved and divergent mechanisms in left-right axis formation. *Genes Dev.* 14, 763–776 (2000).

2. Capdevila, J., Vogan, K. J., Tabin, C. J. & Belmonte, J. C. I. Mechanisms of left-right determination in vertebrates. *Cell.* 101, 9–21 (2000).

3. Hamada, H., Meno, C., Watanabe, D. & Saijoh, Y. Establishment of vertebrate left-right asymmetry. *Nat. Rev. Genet.* 3, 103–13 (2002).

4. Mercola, M. & Levin, M. Left-right asymmetry determination in vertebrates. *Annu. Rev. Cell Dev. Biol.* 17, 779–805 (2001).

5. Rogers, L. J., Vallortigara, G. & Andrew, R. J. Divided brains: The biology and behaviour of brain asymmetries. Cambridge university press, New York. (2013).

6. Schier, A. F. Nodal signalling in vertebrate development. *Annu. Rev. Cell Dev. Biol.* 19,589–621 (2003).

7. Levin, M. Left-right asymmetry in embryonic development: a comprehensive review. *Mech. Dev.* 122, 3–25 (2005).

8. Concha, M. L., Burdine, R. D., Russell, C., Schier, A. F. & Wilson, S. W. A nodal signaling pathway regulates the laterality of neuroanatomical asymmetries in the zebrafish forebrain.*Neuron.* 28, 399–409 (2000).

9. Halpern, M. E., Liang, J. O. & Gamse, J. T. Leaning to the left: laterality in the zebrafish forebrain. *Trends Neurosci.* 26, 308–313 (2003).

10. Ramsdell, A. F. & Yost, H. J. Molecular mechanisms of vertebrate left-right development.*Trends in Genetics.* 14, 459–465 (1998).

11. Zhu, L., Marvin, M. J., Gardiner, A., Lassar, A. B., Mercola, M., Stern, C. D. & Levin, M.Cerverus regulates left–right asymmetry of the embryonic head and heart. *Curr. Biol.* 9,931–938 (1999).

12. Ververs, I. A. P., de Vries, J. I. P., van Geijn, H. P. & Hopkins, B. Prenatal head position from 12–38 weeks. I. Developmental aspects. *Early Hum. Dev.* 39, 83–91 (1994).

13. Previc, F. H. A General theory concerning the prenatal origins of cerebral lateralization in humans. *Psychol. Rev.* 98, 299–334 (1991).

14. Rogers, L. J. Light experience and asymmetry of brain function in chickens. *Nature.* 297,223–225 (1982).

15. Güntürkün, O. Adult persistence of head-turning asymmetry. *Nature.* 421, 711 (2003).

16. Kuo, Z. Y. Ontogeny of embryonic behavior in aves. III. The structural and environmental factors in embryonic behavior. *J. Comp. Psychol.* 13, 245–271 (1932).

17. Rogers, L. J. & Deng, C. Light experience and lateralization of the two visual pathways in the chick. *Behav. Brain Res.* 98, 277–287 (1999).

18. Deng, C. & Rogers, L. J. Social recognition and approach in the chick: Lateralization and effect of visual experience. *Anim. Behav.* 63, 697–706 (2002).

19. Skiba, M., Diekamp, B. & Güntürkün, O. Embryonic light stimulation induces different asymmetries in visuoperceptual and visuomotor pathways of pigeons. *Behav. Brain Res.* 134,149–156 (2002).

20. Mey, J. & Thanos, S. Development of the visual system of the chick I. Cell differentiation and histogenesis. *Brain Res. Brain Res. Rev.* 32, 343–79 (2000).

21. Rogers, L. J. Early experiential effects on laterality: Research on chicks has relevance to other species. *Laterality.* 2, 199–219 (1997).

22. Dharmaretnam, M. & Rogers, L. J. Hemispheric specialization and dual processing in strongly versus weakly lateralized chicks. *Behav. Brain Res.* 162, 62–70 (2005).

23. Rogers, L. J. Light input and the reversal of functional lateralisation in the chicken brain.*Behav. Brain Res.* 38, 211–221 (1990).

24. Manns, M. & Güntürkün, O. Development of the retinotectal system in the pigeon: a choleratoxin study. *Anat. Embryol.* 195, 539–555 (1997).

25. Manns, M. & Güntürkün, O. Monocular deprivation alters the direction of functional and morphological asymmetries in pigeon's visual system. *Behav. Neurosci.* 113, 1–10 (1999).

26. Chiandetti, C. Pseudoneglect and embryonic light stimulation in the avian brain. *Behav. Neurosci.* 125, 775–782 (2011).

27. Chiandetti, C., Regolin, L., Rogers, L. J. & Vallortigara, G. Effects of light stimulation of embryos on the use of position-specific and object-specific cues in binocular and monocular domestic chicks (*Gallus gallus*). *Behav. Brain Res.* 163, 10–17 (2005).

28. Manns, M. & Römling, J. The impact of asymmetrical light input on cerebral hemispheric specialization and interhemispheric cooperation. *Nat. Commun.* 3, 1–5 (2012).

29. Mascetti, G. G. & Vallortigara, G. Why do birds sleep with one eye open? Light exposure of chick embryo as a determinant of monocular sleep. *Curr. Biol.* 11, 971–974 (2001).

30. Andrew, R. J., Johnston, A. N. B., Robins, A. & Rogers, L. J. Light experience and the development of behavioural lateralization in chicks. II. Choice of familiar versus unfamiliar model social partner. *Behav. Brain Res.* 155, 67–76 (2004).

31. Horn, G. & Johnson, M. H. Memory systems in the chick: Dissociations and neuronal analysis.*Neuropsychol.* 27, 1–22 (1989).

32. Johnston, A. N. B. & Rogers, L. J. Light exposure of chick embryo influences lateralized recall of imprinting memory. *Behav. Neurosci.* 113, 1267–1273 (1999).

33. Barth, K. A., Miklósi, A., Watkins, J., Bianco, I. H., Wilson, S. W. & Andrew, R. J. fsi zebrafish show concordant reversal of laterality of viscera, neuroanatomy, and a subset of behavioral responses. *Curr. Biol.* 15, 844–850 (2005).

34. Kennedy, D. N., O'Craven, K. M., Ticho, B. S., Goldstein, A. M., Makris, N. & Henson, J. W.Structural and functional brain asymmetries in human situs inversus. *Neurology.* 53,1260–1265 (1999).

35. McManus, I. C., Martin, N., Stubbings, G. F., Chung, E. M. K. & Mitchison, H. M. Handedness and situs inversus in primary ciliary dyskinesia. *Proc. R. Soc. Lond. B. Biol. Sci.* 271,2579–2582 (2004).

36. Andrew, R. J., Osorio, D. & Budaev, S. Light during embryonic development modulates patterns of lateralisation strongly and similarly in both zebrafish and chick. *Phil. Trans. R. Soc.* 364, 983–989 (2009).

37. Budaev, S. V. & Andrew, R. J. Shyness and behavioural asymmetries in larval zebrafish (*Brachydanio rerio*) developed in light and dark. *Behaviour.* 146, 1037–1052 (2009).

38. Budaev, S. V. & Andrew, R. J. Patterns of early embryonic light exposure determine behavioural asymmetries in zebrafish: A habenular hypothesis. *Behav. Brain Res.* 200, 91–94(2009).

39. Kuan, Y. S., Yu, H. H., Moens, C. B. & Halpern, M. E. Neuropilin asymmetry mediates a left-right difference in habenular connectivity. *Development.* 134, 857–865 (2007).

40. Andrew, R. J. Origins of asymmetry in the CNS. *Semin. Cell Dev. Biol.* 20, 485–490 (2009).

41. Hamburger, V. & Hamilton, H. L. A series of normal stages in the development of the chick embryo. *J. Morphol.* 88, 49–92 (1951).

42. Vallortigara, G., Chiandetti, C. & Sovrano, V. A. Brain asymmetry (animal). *WIREs Cogn. Sci.*2, 146–157 (2011).

43. Ekström, P., Borg, B. & van Veen, T. Ontogenetic development of the pineal organ, parapineal organ, and retina of the three-spined stickleback, *Gasterosteus aculeatus* L. (Teleostei). Development of photoreceptors. *Cell Tissue Res.* 233, 593–609 (1983).

44. Omura, Y. & Oguri, M. Early development of the pineal photoreceptors prior to the retinal differentiation in the embryonic rainbow trout, Oncorhynchus mykiss (Teleostei). *Arch. Histol. Cytol.* 56, 283–91 (1993).

45. Östholm, T., Brännäs, E. & van Veen, T. The pineal organ is the first differentiated light receptor in the embryonic salmon, *Salmo salar* L. *Cell Tissue Res.* 249, 641–646 (1987).

46. Tomonari, S. Takagi, A. Akamatsu, S. Noji, S. & Ohuchi, H. A non-canonical photopigment, melanopsin, is expressed in the differentiating ganglion, horizontal, and bipolar cells of the chicken retina. *Dev. Dynamics.* 234, 783–790 (2005).

47. Diekamp, B., Regolin, L., Güntürkün, O. & Vallortigara, G. A left-sided visuospatial bias in birds. *Curr. Biol.* 15, R372–373 (2005).

expression difference of this gene was investigated between broody and non-broody chickens.

METHODS

Chicken Populations

A total of 24 unrelated chickens were used to identify the mutations in the *DRD1* gene. They were from 6 populations (4 from each) including Red Jungle Fowls (RJF), Taihe Silkies (TS), Xinghua chickens (XH), Gushi chickens (GS), White Recessive Rock Broilers (WRR), and Leghorn Layers (LH). The detail information of the populations was shown in Table 1.

Table 1: The characterization of the populations used in this study

Populations	Origin	Production performance
Red Jungle Fowls (RJF)	Linshan County, Guangxi, China	Seasonal reproduction and broodiness; an egg-production of 60 per year.
Taihe Silkies (TS)	Taihe County, Jiangxi, China	A 70 to 80% incidence of broodiness; an egg-production of 70-80 per year.
Xinghua chickens (XH)	Fengkai County, Guangdong, China	A 70 to 80% incidence of broodiness; an egg-production of 60-90 per year.
Gushi chickens (GS)	Gushi County, Henan, China	A 10 to 20% incidence of broodiness; an egg-production of 141 per year.
White Recessive Rock Broilers (WRR)	Commercial broiler line imported from Kabir Co Ltd, Italy	No broodiness in cage; an egg-production of 180 per year.
Leghorn Layers (LH)	Commercial layer line derived from Italy	No broodiness; an egg-production of 250-300 per year.
Ningdu Sanhuang chickens (NDH)	Ningdu County, Jiangxi, China	A 50 to 60% incidence of broodiness; an egg-production of 110-130 per year.

Xu et al. BMC Genetics 2010 11:17, doi:10.1186/1471-2156-11-17

The population for association study consisted of 644 female Ningdu Sanhuang (NDH) chickens obtained from Guangdong Wens Foodstuff Corporation Ltd. (Guangdong, China). These birds were randomly selected from 1477 unrelated birds. All NDH female chickens were fed with free access to water and feed to 77 d of age, and then changed to feed a corn-soy-bean-based diet with 15% CP and 2,900 kcal of ME/kg. All of them were exposed to a continuous 24 h photoperiod during the first 2 d of age, and then changed to and maintained under a daily light period of 16 h. They were reared in individual laying cages after 90 d of age. In this population, age of first egg (AFE), total egg number from 90 to 300 d of age (EN), total number of qualified eggs from 90 to 300 d of age (QEN), total number of oafish eggs from 90 to 300 d of age (OEN), and weight of first egg (EW) were observed. Qualified eggs were recognized with the criteria as: clean and smooth surface, ellipse shape with a big end and a small end, hard and complete eggshell, stable equilibrium color, similar in size and shape, a good sense of heaviness in hand, crisp noise and hardly breakup after mutual collision. Oafish eggs were abnormal eggs including double-yolk eggs, soft-shell eggs, ruptured eggs, rough-shell eggs, crack eggs, wrinkle eggs and so on.

In addition, from 90 d to 300 d of age the incubation behavior of chickens was observed and recorded at 16:00 pm everyday. The criteria for broody behaviors have been published elsewhere [8]. Briefly, when hens exhibited increased body temperature, nesting, incubating, feather loosening, lacking of luster throughout the body, specific clucking, being more defensive and aggressive, and lost their appetite, they were considered to be in broody. In the association analysis, two parameters, duration of broodiness (DB) and broody frequency (%), were investigated. DB was estimated by the total number of days a hen being in broody during the observation period. Broody frequency (%) was calculated by the percentage of broody chickens, and here individuals exhibiting obvious broody behavior for more than 1 d were identified as broody chickens considering enough sample numbers in statistics.

The distribution pattern of the *DRD1* mRNA was studied in NDH female chickens. The expression differences were compared between 6 NDH chickens in broody and 6 individuals in non-broody in various kinds of tissues (heart, liver, spleen, lung, kidney, breast muscle, leg muscle, gizzard, glandular stomach, pituitary, hypothalamus, ovary, oviduct, duodenum, subcutaneous fat, and abdominal fat). All the tissues of broody chickens were taken at the midpoint (the fourth day after the onset of broody behavior) of broodiness and those of non-broody chickens were taken at the same day. All animal experiments were conducted in accordance with Law of the People's Republic of China on Animal Protection.

DNA Extraction, PCR Amplification and Polymorphism Identification

Genomic DNA was isolated from blood using the traditional method. Two pairs of primers (P1 and P2, shown in Table 2) used for the amplification of the chicken *DRD1* gene were designed according to the published mRNA sequence [GenBank: NM_001144848] by Genetool software (http://www.biologysoft.com/; BioTools, Alberta, Canada). The polymorphisms of the whole *DRD1* coding region were identified through the amplification of a 1970-bp fragment by primer P2 (Table 2). The PCR reactions were carried out in a total volume of 25 µL containing 50 ng of genomic DNA, 1 µM of each primer, 200 µM dNTP, 1.5 mM $MgCl_2$, 1× PCR buffer and 1 U of Taq DNA polymerase (Sangon Biological Engineering Technology Company, Shanghai, China). The amplification was performed in a Eppendorf Mastercycler (Eppendorf Limited, Hamburg, Germany) under the following conditions: 94°C for 3 min; 35 cycles of 94°C for 30 s, n°C for 45 s and 72°C for 1 min; and 72°C for 10 min. PCR products were subjected to a 1% agarose gel electrophoresis and visualized in TFM-40 Ultraviolet Transilluminator (UVP Company, Cambridge, UK) by ethidium bromide staining. Subsequently, the DNA bands were excised from the gel and purified with OMEGA Gel Extraction Kit (OMEGA Bio-Tek Inc., GA, American), and then subcloned into the pMD18-T vector (TaKaRa Biotechnology Co., Ltd., Dalian, China). DNA sequencing was performed by the dideoxy chain-termination method using dye terminator cycle sequencing on Applied Biosystem model 3730 sequencer. The analysis of sequences was conducted by the software DNASTAR V 3.0 (http://www.biologysoft.com; Steve ShearDown, 1998-2001 version reserved by DNASTAR Inc., Madison, Wisconsin, USA).

Table 2: Detail information for primers of the chicken *DRD1* gene

Primer name	Primer sequence (5'→3')	Length¹ (bp)	Location²	AT³ (°C)	Genotyping method
P1	F:CCGGTGAGTACCCTGCTTT R:GTGCTTTTCCTCTGCTTTGG	1504	-1961 ~ -458	59	/
P2	F:AGTGAAGAATTGCTCGCTGA R:GGTTTTGCTGGGTACACCTT	1970	-589 ~ +1381	57	/
P3	F:CACTATGGATGGGAAGGGTTG R: GGCCACCCAGATGTTGCAAAATG	283	G+123A T+198C	62	*BseNI* *cfrl*
P4	F: CAGCCCATTCAGGTACGAGAGGA R: ATTCGACTCTTTGGGGCTGGAC	793	A+505G C+765T C+1011T G+1065A C+1107T	65.5	sequencing
P5	F: TTTCCTTCATCCCCGTGCAGCT R: GCTGCTTCTGTTGCCACTTGTGT	290	+458 ~ +747	63	/

| P6 (actin) | F: CCCCAAAGCCAACAGAGAGA | 274 | / | 63 | / |
| | R: GGTGGTGAAGCTGTAGCCTCTC | | | | |

[1]The length of PCR products; [2]referred to the locations in the *DRD1* gene, the first nucleotide of translation start codon was designated as +1. [3]indicated annealing temperature.

Xu et al. *BMC Genetics* 2010 **11**:17, doi:10.1186/1471-2156-11-17

Genotyping of Polymorphisms

Primers (P3 and P4) used for genotyping of polymorphisms in the *DRD1* coding region were described in Table 2. Genotypes of G+123A and T+198C were determined with PCR-RFLP method using genomic DNA from the 644 NDH individuals as templates. PCR products were subjected to digestion for 16 h at 37°C with the restriction enzyme BseNI and CfrI, respectively. The digestion mixture was composed of 8 μL PCR products, 1 × digestion buffer, and 3.0 U of each enzyme. Subsequently the fragments were visualized by TFM-40 Ultraviolet Transilluminator (UVP Company, Cambridge, UK) following separation in 2.5% agarose gels and staining with ethidium bromide. For the other SNPs in the coding region, genotyping was carried out by direct sequencing.

RNA Extraction and cDNA Synthesis

Twelve NDH chickens, 6 broody individuals and 6 non-broody ones, were used in expression analysis. A total of 16 tissues, including heart, liver, spleen, lung, kidney, breast muscle, leg muscle, gizzard, glandular stomach, pituitary, hypothalamus, ovary, oviduct, duodenum, subcutaneous fat, and abdominal fat, were collected from each chicken. Among the 16 tissues, six tissues of pituitary, hypothalamus, ovary, oviduct, subcutaneous fat, and abdominal fat were important parts of chicken reproduction physiology system. Whereas other tissues including heart, liver, spleen, lung, kidney, breast muscle, leg muscle, gizzard, glandular stomach and duodenum were chosen to be the background tissues in chicken broodiness research. The dissected tissues were frozen in liquid nitrogen immediately and subsequently stored at -80°C until used. Total RNA was extracted with TRIzol reagent (Invitrogen, Carlsbad, CA, USA) following the manufacturer's instructions and then treated with DNase (Promega, Madison, WI, USA). The DNase reaction were composed of 1 μg of total RNA, 1 U RNase-free DNase, 1 μL 10 × Reaction Buffer and 7 μL nuclease-free water. The mixture was incubated at 37°C for 30 min followed by denaturation at 65°C for 10 min and snap cooled on ice for 2 min. The quality and purity of the RNA were checked by agarose gel electrophoresis and spectrophotometry. cDNA was synthesized in a final volume of 20 μL including 1 μg of total RNA, 1× MMLV Buffer, 1 mM of each dNTPs, 2.5 μM oligo $(dT)_{18}$, 0.5 μL (40 U/μl) RNase inhibitor, 100 U MMLV SuperScript III reverse transcriptase (Invitrogen, Carlsbad, CA, USA). The reverse transcription was processed for 40 min at 42°C followed by heating for 5 min at 95°C and cooling on ice.

Quantitative Real-Time PCR

Quantitative real-time PCR (qPCR) was performed with the ABI PRISM 7000 sequence detection system (Applied Biosystems, Foster City, CA, USA) using the SYBR Green PCR Master Mix. The obtained cDNAs were used as templates for qPCR amplification. Primers used for the qPCR were designed by Primer Express 2.0 software (Applied Biosystems, Foster City, CA, USA). A housekeeping gene, chicken *β-actin* gene [GenBank: LOC396526], was used as internal control. Therefore, two sets of primers (shown in Table 2), P5 and P6, were designed and used for the qPCR amplification of chicken *DRD1* gene and chicken *β-actin* gene, respectively. Each reaction mixture contained 10 μL of SYBR Green PCR Master Mix, 2 μL of each primer (10 μM), 4 μL ultrapure RNase-free water and 2 μL of cDNA in a final volume of 20 μL. Standard amplification conditions were as follows: 95°C for 3 min; 40 cycles of 95°C for 30 s, 63°C for 30 s and 72°C for 40 s. Fluorescent signal were collected after the extension at 72°C in each cycle. Amplification of DRD1 and β-actin for each sample was run simultaneously in separate tubes and in duplicates. A negative control with sterile water as template was run for each primer in order to control the reagent contamination. The whole experiment was repeated at least twice. After amplification, dissociation curve analysis was conducted to ensure only one product. And then the product was sequenced to confirm amplification of the correct sequences.

Statistical Analysis

Identification of the Chicken DRD1 Gene Polymorphisms and Prediction of the Transcription Factor Binding Sites in the 5' Flanking Region

DNAMAN (Lynnon Biosoft) was used for DNA contig assembly, sequence editing, and sequence translation. The identification of mutated sites was performed by MegAlign program of DNASTAR software (http://www.biologysoft.com/; Steve ShearDown, 1998-2001 version reserved by DNASTAR Inc., Madison, Wisconsin, USA). The potential transcription factor binding sites of the 5'-flanking region polymorphisms were predicted by two bioinformatic websites ofhttp://motif.genome.jp and http://www.gene-regulation.com/pub/programs/alibaba2 following the setting parameters. The same results identified by the two websites were finally chosen.

Haplotype Inference and Marker -Trait Association Analysis

Hardy-Weinberg's equilibrium and the haplotype structure were analyzed by Haploview version 3.32 software http://www.broad.mit.edu/mpg/haploview/) [37]. Haplotypes were inferred based on the haplotype structure by the PHASE 2.0 softwarehttp://www.stat.washington.edu/stephens/software.html [38].

Association analysis of polymorphisms or haplotypes with egg production and broodiness traits were conducted by SAS GLM procedure (SAS Institute Inc., Cary, NC, USA) using the following model:

$$Y_{ij} = \mu + G_i + H_j\, e_{ij}$$

Where Y_{ij} is an observation on the traits, μ is the overall population mean, G_i is the effect of genotype, H_j is the fixed effect of hatch and the e_{ij} is the residual error. Multiple comparisons were performed with least squares means using the following procedure:

$$Y_i - \overline{Y}_i = \left(Y_i - \hat{Y}_i\right) + \left(\hat{Y}_i - \overline{Y}_i\right)$$

Where $\sum\left(Y_i - \hat{Y}_i\right)^2$ is the least value, and $\sum\left(Y_i - \hat{Y}_i\right)^2 = 0$. The results were presented as least square means ± standard error.

The comparisons of broody frequency among different genotypes or diplotypes in each site were evaluated by chi-square (χ^2) tests performed on a 2 × 3 (or n) contingency table. A $P \leq 0.05$ was considered statistically significant in all analysis.

Expression Analysis of DRD1 mRNA

Table 5: The corresponding base combinations for two haplotype blocks

Block 1 Haplotype	G+123A	T+198C	Frequency	Block 2 Haplotype	G+1065A	C+1107T	Frequency
H1(AT)	A	T	0.0079	E1(AT)	A	T	0.0016
H2(AC)	A	C	0.2128	E2(AC)	A	C	0.5747
H3(GT)	G	T	0.7049	E3(GT)	G	T	0.1826
H4(GC)	G	C	0.0744	E4(GC)	G	C	0.2411

Xu et al. BMC Genetics 2010 **11**:17, doi:10.1186/1471-2156-11-17

Association of Polymorphisms in the DRD1 Coding Region with Chicken Egg Production and Broodiness Traits

The G+123A was significantly associated with chicken broody frequency (P < 0.05). Furthermore, the EN values of chickens with the AA genotype were significantly higher than those with the GG genotype (P < 0.05) (Table 6). C+1107T was in significant association (P < 0.05) with chicken broody frequency (Table 7). No significant association was found in the other 4 markers (A-179T, T+198C, C+765T, and G+1065A) with chicken egg production and broodiness traits (P > 0.05).

Table 6: Association of the G+123A with egg production traits and broody traits in Ningdu Sanhuang Chickens

Traits[1]	P value	GG[2] (385)	AG[2] (217)	AA[2] (31)
AFE(d)	0.58	135.7 ± 0.6[a]	136.3 ± 0.7[a]	137.3 ± 1.9[a]
EN	0.10	113.3 ± 1.5[a]	114.6 ± 1.9[ab]	124.2 ± 4.9[b]
QEN	0.17	109.8 ± 1.5[a]	110.3 ± 1.8[a]	119.2 ± 4.8[a]
OEN	0.14	3.6 ± 0.3[a]	4.3 ± 0.4[a]	5.0 ± 1.0[a]
EW (g)	0.89	45.8 ± 0.2[a]	45.7 ± 0.3[a]	45.5 ± 0.6[a]
DB(d)	0.35	7.9 ± 0.8[a]	7.0 ± 1.0[a]	4.4 ± 2.5[a]
Number of nonbroody chickens	/	186	116	22
Number of broody chickens	/	199	101	9
Broody frequency (%)	/	51.69	46.54	29.03
χ^2value	< 0.05	6.58*		

[1]AFE = age of first egg; EN = total egg number from 90 to 300 d of age; QEN = total number of qualified eggs from 90 to 300 d of age; OEN = total number of oafish eggs from 90 to 300 d of age; EW = weight of first egg; DB = duration days of broodiness.[2]Least-square means ± standard errors (SE); Number in brackets referred to the number of tested chickens of each genotype. [a, b]means within a row with no

common superscript are different significantly (P < 0.05). * indicated P < 0.05. $\chi^2_{0.05}$(df = 2) = 5.99.

Xu et al. BMC Genetics 2010 11:17, doi:10.1186/1471-2156-11-17

Table 7: Association of the C+1107T with egg production traits and broody traits in Ningdu Sanhuang Chickens

Traits[1]	P value	CC[2] (403)	TC[2] (201)	TT[2] (13)
AFE(d)	0.24	136.0 ± 0.6	135.7 ± 0.7	140.7 ± 2.9
EN	0.27	113.6 ± 1.5	115.1 ± 2.0	125.5 ± 7.6
QEN	0.36	109.8 ± 1.4	110.9 ± 1.9	120.3 ± 7.4
OEN	0.37	3.7 ± 0.3	4.2 ± 0.4	5.3 ± 1.5
EW (g)	0.88	45.8 ± 0.2	45.8 ± 0.3	45.3 ± 0.9
Duration of broodiness (d)	0.85	7.5 ± 0.7	7.3 ± 1.0	5.3 ± 3.9
Number of nonbroody chickens	/	195	108	11
Number of broody chickens	/	208	93	2
Broody frequency (%)	/	51.61	46.27	15.38
χ^2value	< 0.05	7.58*		

[1]AFE = age of first egg; EN = total egg number from 90 to 300 d of age; QEN = total number of qualified eggs from 90 to 300 d of age; OEN = total number of oafish eggs from 90 to 300 d of age; EW = weight of first egg; DB = duration days of broodiness.[2]Least-square means ± standard errors (SE); Number in brackets referred to the number of tested chickens of each genotype. [a, b]means within a row with no common superscript are different significantly (P < 0.05). * indicated P < 0.05. $\chi^2_{0.05}$(df = 2) = 5.99.

Xu et al. BMC Genetics 2010 11:17, doi:10.1186/1471-2156-11-17

Association of the Haplotypes with Chicken Egg Production and Broodiness Traits

A total of 623 individuals with 6 diplotypes (30 of H2H2, 178 of H2H3, 31 of H2H4, 324 of H3H3, 57 of H3H4, 3 of H4H4) were used in association analysis in the block 1. Significant association (P = 0.03) of the haplotypes of G+123A and T+198C with EW was observed. H2H4 had much lower value of EW (mean = 44.1 g) and was highly significantly different (P < 0.01) from H2H3, significantly (P < 0.05) from H3H4. Nevertheless, H2H2 had much higher value of EN (mean = 124.0) and QEN (mean = 118.9) and was significantly different from H3H3 (P = 0.04) and H4H4 (P = 0.02), respectively.

A total of 614 individuals with 6 diplotypes (200 of E2E2, 134 of E2E3, 174 of E2E4, 11 of E3E3, 67 of E3E4, 28 of E4E4) were used in association analysis in the block 2. The haplotypes of G+1065A and C+1107T were significantly associated (χ^2 value $_{(df = 5)}$ = 11.08, 0.01 < P < 0.05) with broody frequency. Nevertheless, E3E3 had much higher value of AFE (mean = 142.2) than other diplotypes and was significantly different with E2E2, E2E3, and E3E4 (P < 0.05).

Tissue-Specific Expression of the DRD1 and the mRNA Comparison between Broodiness and Non-Broodiness Chickens

The *DRD1* mRNA was differentially expressed in distinct tissues (Figure 1). There was almost no mRNA present in gizzard. Low mRNA level was observed in liver, spleen, lung, breast muscle, leg muscle, as well as in duodenum. Instead, much higher levels of the *DRD1* expression were detected in tissues such as heart, kidney, oviduct, glandular stomach, hypothalamus, and pituitary. Remarkably, the highest levels of the *DRD1* expression were found in subcutaneous fat of non-broodiness chickens, and then abdominal fat.

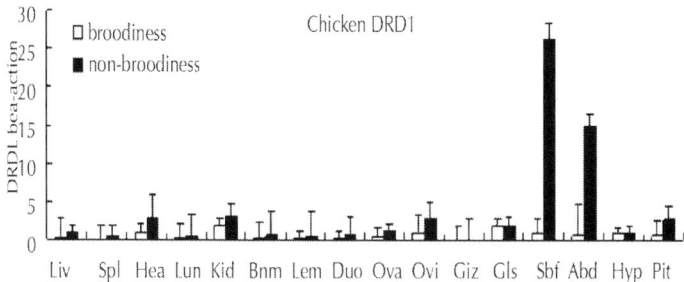

Figure 1: The distribution of *DRD1* mRNA in broodiness and non-broodiness chickens. The horizontal axis and vertical axis indicate different tissues and $2^{-\Delta\Delta Ct}$ value (mean ± SEM), respectively. Liv = liver, Spl = spleen, Hea = heart; Lun = lung, Kid = kidney, Brm = breast muscle, Lem = leg muscle, Duo = duodenum, Ova = ovary, Ovi = oviduct, Giz = gizzard, Gls= glandular stomach, Sbf = subcutaneous fat, Abd = abdominal fat, Hyp = hypothalamus, Pit = pituitary.

In subcutaneous fat and abdominal fat, significantly difference of the *DRD1* mRNA was found between broodiness and non-broodiness chickens (P < 0.05). The level of non-broodiness was 26 to 28 times higher than that of broodiness. The expression of non-broodiness was observed to be 5-fold greater than that of broodiness in pituitary. Also a prominent decrease from non-broodiness to broodiness was displayed in heart, oviduct, and kidney. In these tissues, the expression of non-broodiness was 2-3 times higher as compared to that of broodiness. The level of glandular stomach seen in non-broodiness was almost the same as in broodiness. Similarly, the same level was found in non-broodiness and broodiness hypothalamus (Figure 1).

DISCUSSION

In this study, abundant polymorphisms were found in the chicken *DRD1* gene. Soller reported that SNP frequencies in poultry species ranged from 1:48 to 1:1632 bp [40]. Here SNP frequency of the chicken *DRD1* gene was 1:115 bp and it was similar to previous study [41]. In this study, the absence of TATA and CAAT boxes was found in chicken *DRD1* promoter, as reported in human [42]. Mutations in the promoter region can cause changes of transcription factor binding sites and consequently may affect the transcription and the phenotype [43]. In the 5′ flanking region of the chicken *DRD1* gene, there were multiple putative binding sites for transcription factor Sp1 and consensus sequences for AP1 and AP2 binding sites. It was consistent with the analysis of human*DRD1* gene [44].

A variety of structural variations occurred in different regions of G protein coupled receptor proteins have been found to be related with diseases [45,46]. In the present study, one non-synonymous mutation (Ser169Gly) was present in the extracellular domain. As no polymorphism of this site was observed in the NDH population, its effects on chicken egg production and broodiness still required further study in other populations.

SNP in cytoplasmic tail and transmembrane I seemed to have great effects on egg production and broodiness. In this study, three mutations found within the cytoplasmic tail of the chicken *DRD1*gene might cause the change of various functions even if they were synonymous. C+1107T, a mutation located in the cytoplasmic tail, was found to be associated with chicken broodiness, and haplotype analysis also provided similar results. In addition, G+123A, a variation in transmembrane I of the *DRD1* gene, was associated with chicken egg production and broodiness traits. The polymorphisms of G+123A and C+1107T may be acted as Marker assistant selection (MAS) markers of reducing incidence of broodiness and improving egg production in modern poultry industry. In the same NDH population of our former study, two SNPs of the chicken *DRD2* gene, A-16105G and T+619C, were also found to be significantly associated with broody frequency and duration of broodiness, respectively [47].

Other studies also indicated that the cytoplasmic tail of D1-like receptors, especially the N-terminal segment termed as the fourth intracellular loop, played a crucial role in the regulation of the activation of adenylyl cyclase, ligand binding, expression, and G protein coupling properties [48-52]. Members of G protein coupled receptors displayed considerable amino acid sequence conservation within transmembrane domains [53]. Through bioinformatics analysis, the presumed transmembrane domains of the *DRD1* gene were proved to be highly conserved in diverse species. Many previous studies reported that some mutants present in transmembrane domains of the Dl receptor affect ligand interactions and receptor signal transduction [54-56]. It seemed that the variation may play a crucial role on egg production and broodiness traits by affecting ligand binding or signal transduction.

In mammals, the DRD1 gene was found to express in the tissues of striatum, nucleus accumbens, cerebral cortex, amygdale, olfactory tubercle, retina, limbic system, hypothalamus, and thalamus[57,58], but not in cerebellum, hippocampus, mesencephalon, pituitary, kidney, liver, lung and heart tissues [42,59,60]. In avian, it was revealed that the distribution of the DRD1 gene in the forebrain was substantially similar to that of mammals [61]. In this study, similar to mammals, high level of the chicken DRD1 mRNA was detected in the hypothalamus. However, high to moderateDRD1 mRNA were also detected in chicken heart, kidney, oviduct, glandular stomach, hypothalamus, pituitary, and adipose tissues and a considerably low but still detectable expression was found in liver, spleen, lung, muscle, and duodenum.

The level of the DRD1 mRNA was quantified in the brain of the domestic turkey hen during the reproductive cycle and it was expressed throughout the hypothalamus and pituitary [35]. But no significant difference of DRD1 mRNA abundance was observed in hypothalamic and pituitary throughout the reproductive cycle. Similarly, this study suggested that in hypothalamus, as well as in glandular stomach, the DRD1 mRNA levels seen in broody chickens were essentially the same as in non-broody ones. However, higher DRD1 mRNA content was found in pituitary (5-fold), heart (3-fold), oviduct (2.8-fold), and kidney (2-fold) of non-broody hens as compared to that of incubating hyperprolactinemic hens. In particular, it was interesting that there was a dramatic expression difference in adipose tissues from non-broodiness to broodiness stage. The level of the DRD1mRNA in non-broody chickens was 26 to 28 times greater than that of broody chickens in adipose tissues including subcutaneous fat and abdominal fat. As chickens in broodiness had lower fatty content compared with non-broody chickens, the subtle decreased mRNA in broody chickens suggested that DRD1 was probably involved in fat deposition. In general, a high abundance ofDRD1 mRNA was found in non-broodiness compared with broodiness in each tissue, except for the glandular stomach and hypothalamus. All these findings indicated that the DRD1 gene was probably related to chicken broodiness.

CONCLUSIONS

In summary, the results of association analysis and the expression comparison of broody chickens with non-broody chickens demonstrated that the *DRD1* had important effects on chicken egg production and broodiness incidence.

AUTHORS' CONTRIBUTIONS

HX carried out the mRNA research, analyzed the data and drafted the manuscript. XS contributed to the genotyping of most of the SNP, MZ and MF participated in the data analyses. HZ contributed to materials collection. QN and XZ contributed to the design of the study, the supervision of the study and the revision of this manuscript. All authors read and approved the final manuscript.

ACKNOWLEDGEMENTS

This study was supported by a grant from the National Natural Scientific Foundation of China, project no. 30471241, a grant from the Major State Basic Research Development Program of China, project no. 2006CB102107 and a grant from the National High Technology Research and Development Program of China (863 Program), project no. 2006AA10A120.

REFERENCES

1.   Sharp PJ, Dawson A, Lea RW: Control of luteinizing hormone and prolactin secretion in birds. *Comp Biochem Physiol C Pharmacol Toxicol Endocrinol* 1998, 119:275-282.

2.   Sharp PJ: Genes for persistency of egg laying: White Leghorns and broodiness. *Roslin Institute Edinburgh Annual Report* 2004, 38-42.

3.   Romanov MN, Talbot RT, Wilson PW, Sharp PJ: Inheritance of broodiness in the domestic fowl. *Br Poult Sci* 1999, 40(Suppl):20-21.

4. Romanov MN, Talbot RT, Wilson PW, Sharp PJ: Genetic control of incubation behavior in the domestic hen. *Poult Sci* 2002, 81:928-931.

5. Cui JX, Du HL, Liang Y, Deng XM, Li N, Zhang XQ: Association of polymorphisms in the promoter region of chicken prolactin with egg production. *Poult Sci* 2006, 85:26-31.

6. Jiang RS, Xu GY, Zhang XQ, Yang N: Association of polymorphisms for prolactin and prolactin receptor genes with broody traits in chickens. *Poult Sci* 2005, 84:839-845.

7. Liang Y, Cui J, Yang G, Leung FC, Zhang X: Polymorphisms of 5' flanking region of chicken prolactin gene. *Domest Anim Endocrinol* 2006, 30:1-16.

8. Zhou M, Lei M, Rao Y, Nie Q, Zeng H, Xia M, Liang F, Zhang D, Zhang X: Polymorphisms of vasoactive intestinal peptide receptor-1 gene and their genetic effects on broodiness in chickens. *Poult Sci* 2008, 87:893-903.

9. Sharp PJ, Macnamee MC, Sterling RJ, Lea RW, Pedersen HC: Relationships between prolactin, LH and broody behaviour in bantam hens. *J Endocrinol* 1988, 118:279-286.

10. el Halawani ME, Silsby JL, Youngren OM, Phillips RE: Exogenous prolactin delays photo-induced sexual maturity and suppresses ovariectomy-induced luteinizing hormone secretion in the turkey (Meleagris gallopavo). *Biol Reprod* 1991, 44:420-424.

11. el Halawani ME, Silsby JL, Behnke EJ, Fehrer SC: Hormonal induction of incubation behavior in ovariectomized female turkeys (Meleagris gallopavo). *Biol Reprod* 1986, 35:59-67.

12. March JB, Sharp PJ, Wilson PW, Sang HM: Effect of active immunization against recombinant-derived chicken prolactin fusion protein on the onset of broodiness and photoinduced egg laying in bantam hens. *J Reprod Fertil* 1994, 101:227-233.

13. Ben-Jonathan N, Hnasko R: Dopamine as a prolactin (PRL) inhibitor. *Endocr Rev* 2001, 22:724-763.

14. Reymond MJ, Porter JC: Involvement of hypothalamic dopamine in the regulation of prolactin secretion. *Horm Res* 1985, 22:142-152.

15. Xu M, Proudman JA, Pitts GR, Wong EA, Foster DN, el Halawani ME: Vasoactive intestinal peptide stimulates prolactin mRNA

expression in turkey pituitary cells: effects of dopaminergic drugs. *Proc Soc Exp Biol Med* 1996, 212:52-62.

16.   Cools R: Role of dopamine in the motivational and cognitive control of behavior. *Neuroscientist* 2008, 14:381-395.

17.   Nieoullon A, Coquerel A: Dopamine: a key regulator to adapt action, emotion, motivation and cognition. *Curr Opin Neurol* 2003, 16(Suppl 2):3-9.

18.   Hansen KA, Zhang Y, Colver R, Tho SP, Plouffe L Jr, McDonough PG: The dopamine receptor D2 genotype is associated with hyperprolactinemia. *Fertil Steril* 2005, 84:711-718.

19.   Missale C, Nash SR, Robinson SW, Jaber M, Caron MG: Dopamine receptors: from structure to function. *Physiol Rev* 1998, 78:189-225.

20.   Kebabian JW, Calne DB: Multiple receptors for dopamine. *Nature* 1979, 277:93-96.

21.   Gingrich JA, Caron MG: Recent advances in the molecular biology of dopamine receptors. *Annu Rev Neurosci* 1993, 16:299-321.

22.   Seeman P, Van Tol HH: Dopamine receptor pharmacology. *Trends Pharmacol Sci* 1994, 15:264-270.

23.   Youngren OM, Pitts GR, Phillips RE, el Halawani ME: The stimulatory and inhibitory effects of dopamine on prolactin secretion in the turkey. *Gen Comp Endocrinol* 1995, 98:111-117.

24.   Porter TE, Grandy D, Bunzow J, Wiles CD, Civelli O, Frawley LS: Evidence that stimulatory dopamine receptors may be involved in the regulation of prolactin secretion. *Endocrinology* 1994, 134:1263-1268.

25.   Youngren OM, Pitts GR, Phillips RE, el Halawani ME: Dopaminergic control of prolactin secretion in the turkey. *Gen Comp Endocrinol* 1996, 104:225-230.

26.   Youngren O, Chaiseha Y, Al-Zailaie K, Whiting S, Kang SW, El Halawani M: Regulation of prolactin secretion by dopamine at the level of the hypothalamus in the turkey. *Neuroendocrinology* 2002, 75:185-192.

27.   Youngren OM, Chaiseha Y, El Halawani ME: Regulation of prolactin secretion by dopamine and vasoactive intestinal peptide at the level of the pituitary in the turkey. *Neuroendocrinology* 1998, 68:319-325.

28. Al Kahtane A, Chaiseha Y, El Halawani M: Dopaminergic regulation of avian prolactin gene transcription. *J Mol Endocrinol* 2003, 31:185-196.

29. Hall TR, Chadwick A: Dopaminergic inhibition of prolactin release from pituitary glands of the domestic fowl incubated in vitro. *J Endocrinol* 1984, 103:63-69.

30. Millam JR, Burke WH, El Halawani ME, Ogren LA: Preventing broodiness in turkey hens with a dopamine receptor blocking agent. *Poult Sci* 1980, 59:1126-1131.

31. Shi ZD, Liang SD, Bi YZ: Studies on the role of hypothalamic dopamine and 5-hydroxytryptamine in regulation of broodiness in chicken hens. *Proceedings of international conference on bird reproduction: 22-24 September 1999; Tours, France*

32. Sartsoongnoen N, Kosonsiriluk S, Prakobsaeng N, Songserm T, Rozenboim I, Halawani ME, Chaiseha Y: The dopaminergic system in the brain of the native Thai chicken, Gallus domesticus: localization and differential expression across the reproductive cycle. *Gen Comp Endocrinol* 2008, 159:107-115.

33. Laitinen JT: Dopamine stimulates K+ efflux in the chick retina via D1 receptors independently of adenylyl cyclase activation. *J Neurochem* 1993, 61:1461-1469.

34. Demchyshyn LL, Sugamori KS, Lee FJ, Hamadanizadeh SA, Niznik HB: The dopamine D1D receptor. Cloning and characterization of three pharmacologically distinct D1-like receptors from Gallus domesticus. *J Biol Chem* 1995, 270:4005-4012.

35. Schnell SA, You S, El Halawani ME: D1 and D2 dopamine receptor messenger ribonucleic acid in brain and pituitary during the reproductive cycle of the turkey hen. *Biol Reprod* 1999, 60:1378-1383.

36. Chaiseha Y, Youngren O, Al-Zailaie K, El Halawani M: Expression of D1 and D2 dopamine receptors in the hypothalamus and pituitary during the turkey reproductive cycle: colocalization with vasoactive intestinal peptide. *Neuroendocrinology* 2003, 77:105-118.

37. Barrett JC, Fry B, Maller J, Daly MJ: Haploview: analysis and visualization of LD and haplotype maps. *Bioinformatics* 2005, 21:263-265.

38. Stephens M, Smith NJ, Donnelly P: A new statistical method for haplotype reconstruction from population data. *Am J Hum Genet* 2001, 68:978 989.

39. Livak KJ, Schmittgen TD: Analysis of relative gene expression data using real-time quantitative PCR and the 2(-Delta Delta C(T)) Method. *Methods* 2001, 25:402-408.

40. Soller M, Weigend S, Romanov MN, Dekkers JC, Lamont SJ: Strategies to assess structural variation in the chicken genome and its associations with biodiversity and biological performance. *Poult Sci* 2006, 85:2061-2078.

41. Wong GK, Liu B, Wang J, Zhang Y, Yang X, Zhang Z, Meng Q, Zhou J, Li D, Zhang J, Ni P, Li S, Ran L, Li H, Zhang J, Li R, Li S, Zheng H, Lin W, Li G, Wang X, Zhao W, Li J, Ye C, Dai M, Ruan J, Zhou Y, Li Y, He X, Zhang Y, Wang J, Huang X, Tong W, Chen J, Ye J, Chen C, Wei N, Li G, Dong L, Lan F, Sun Y, Zhang Z, Yang Z, Yu Y, Huang Y, He D, Xi Y, Wei D, Qi Q, Li W, Shi J, Wang M, Xie F, Wang J, Zhang X, Wang P, Zhao Y, Li N, Yang N, Dong W, Hu S, Zeng C, Zheng W, Hao B, Hillier LW, Yang SP, Warren WC, Wilson RK, Brandström M, Ellegren H, Crooijmans RP, Poel JJ, Bovenhuis H, Groenen MA, Ovcharenko I, Gordon L, Stubbs L, Lucas S, Glavina T, Aerts A, Kaiser P, Rothwell L, Young JR, Rogers S, Walker BA, van Hateren A, Kaufman J, Bumstead N, Lamont SJ, Zhou H, Hocking PM, Morrice D, de Koning DJ, Law A, Bartley N, Burt DW, Hunt H, Cheng HH, Gunnarsson U, Wahlberg P, Andersson L, Kindlund E, Tammi MT, Andersson B, Webber C, Ponting CP, Overton IM, Boardman PE, Tang H, Hubbard SJ, Wilson SA, Yu J, Wang J, Yang H, International Chicken Polymorphism Map Consortium: A genetic variation map for chicken with 2.8 million single-nucleotide polymorphisms. *Nature* 2004, 432:717-722.

42. Vallone D, Picetti R, Borrelli E: Structure and function of dopamine receptors. *Neurosci Biobehav Rev* 2000, 24:125-132.

43. Xu H, Gregory SG, Hauser ER, Stenger JE, Pericak-Vance MA, Vance JM, Züchner S, Hauser MA: SNPselector: a web tool for selecting SNPs for genetic association studies. *Bioinformatics* 2005, 21:4181-4186.

44. Minowa MT, Minowa T, Monsma FJ Jr, Sibley DR, Mouradian MM: Characterization of the 5' flanking region of the human

D1A dopamine receptor gene. *Proc Natl Acad Sci USA* 1992, 89:3045-3049.

45. Neidhardt J, Barthelmes D, Farahmand F, Fleischhauer JC, Berger W: Different amino acid substitutions at the same position in rhodopsin lead to distinct phenotypes. *Invest Ophthalmol Vis Sci* 2006, 47:1630 1635.

46. Schipani E, Kruse K, Jüppner H: A constitutively active mutant PTH-PTHrP receptor in Jansen-type metaphyseal chondrodysplasia. *Science* 1995, 268:98-100.

47. Xu HP, Shen X, Zhou M, Luo CL, Kang L, Liang Y, Zeng H, Zhang DX, Nie QH, Zhang XQ: The dopamine D2 receptor gene polymorphisms associated with chicken broodiness. *Poult Sci* 2010, 89:428-438.

48. Chaar ZY, Jackson A, Tiberi M: The cytoplasmic tail of the D1A receptor subtype: identification of specific domains controlling dopamine cellular responsiveness. *Neurochem* 2001, 79:1047-1058.

49. Jackson A, Iwasiow RM, Tiberi M: Distinct function of the cytoplasmic tail in human D1-like receptor ligand binding and coupling. *FEBS Lett* 2000, 470:183-188.

50. Jensen AA, Pedersen UB, Kiemer A, Din N, Andersen PH: Functional importance of the carboxyl tail cysteine residues in the human D1 dopamine receptor. *J Neurochem* 1995, 65:1325-1331.

51. Tumova K, Iwasiow RM, Tiberi M: Insight into the mechanism of dopamine D1-like receptor activation. Evidence for a molecular interplay between the third extracellular loop and the cytoplasmic tail. *J Biol Chem* 2003, 278:8146-8153.

52. Tumova K, Zhang D, Tiberi M: Role of the fourth intracellular loop of D1-like dopaminergic receptors in conferring subtype-specific signaling properties. *FEBS Lett* 2004, 576:461-467.

53. Probst WC, Snyder LA, Schuster DI, Brosius J, Sealfon SC: Sequence alignment of the G-protein coupled receptor superfamily. *DNA Cell Biol* 1992, 11:1-20.

54. Cho W, Taylor LP, Akil H: Mutagenesis of residues adjacent to transmembrane prolines alters D1 dopamine receptor binding and signal transduction. *Mol Pharmacol* 1996, 50:1338-1345

55. Pollock NJ, Manelli AM, Hutchins CW, Steffey ME, MacKenzie RG, Frail DE: Serine mutations in transmembrane V of the dopamine D1 receptor affect ligand interactions and receptor activation. *J Biol Chem* 1992, 267:17780-17786.

56. Tomic M, Seeman P, George SR, O'Dowd BF: Dopamine D1 receptor mutagenesis: role of amino acids in agonist and antagonist binding. *Biochem Biophys Res Commun* 1993, 191:1020-1027.

57. Fremeau RT Jr, Duncan GE, Fornaretto MG, Dearry A, Gingrich JA, Breese GR, Caron MG:Localization of D1 dopamine receptor mRNA in brain supports a role in cognitive, affective, and neuroendocrine aspects of dopaminergic neurotransmission. *Proc Natl Acad Sci USA* 1991, 88:3772-3776.

58. Jackson DM, Westlind-Danielsson A: Dopamine receptors: molecular biology, biochemistry and behavioural aspects. *Pharmacol Ther* 1994, 64:291-370.

59. Monsma FJ Jr, Mahan LC, McVittie LD, Gerfen CR, Sibley DR: Molecular cloning and expression of a D1 dopamine receptor linked to adenylyl cyclase activation. *Proc Natl Acad Sci USA* 1990, 87:6723-6727.

60. Niznik HB, Van Tol HH: Dopamine receptor genes: new tools for molecular psychiatry. *J Psychiatry Neurosci* 1992, 17:158-180.

61. Schnabel R, Metzger M, Jiang S, Hemmings HC Jr, Greengard P, Braun K: Localization of dopamine D1 receptors and dopaminoceptive neurons in the chick forebrain. *J Comp Neurol* 1997, 388:146-168.

Feeding Behaviour of Broiler Chickens: A Review on the Biomechanical Characteristics

Neves DPI;Banhazi TMII;and Nääs IAI

ISchool of Agriculture Engineering, State University of Campinas, Cândido Rondon Ave, 501, Brazil

IINational Centre for Engineering in Agriculture, Faculty of Engineering and Surveying, University of Southern Queensland, West Street, Toowoomba QLD, 4350, Australia

ABSTRACT

Feed related costs are the main drivers of profitability of commercial poultry farms, and good nutrition is mainly responsible for the exceptional growth rate responses of current poultry species. So far, most research on the poultry feeding behaviour addresses the productivity indices and birds' physiological responses, but few studies have considered the biomechanical characteristics involved in this process. This paper aims to review biomechanical issues related to feed behaviour of domestic chickens to address some issues related to the feed used in commercial broiler chicken production, considering feed particle size, physical form and the impact of feeders during feeding. It is believed that the biomechanical evaluation might suggest a new way for feed processing to meet the natural feeding behaviour of the birds.

INTRODUCTION

The poultry industry is the most dynamic sector within the global meat business during the last decade, with the greatest growth reflected in the food global demand increase. It is expected that, in the next years, the meat industry will increase production driven by global population growth, especially in developing countries. Chickens and turkeys are the most common sources of poultry meat, but there is also commercially available meat from ducks, geese, pigeons, quails, pheasants, ostriches and emus. Consumer preference also has been changing in many developed countries, characterized by greater demand for low-calorie foods and changes in lifestyle, which reduces the consumer time spent on food preparation. By this approach, the chicken meat highlights and the largest producer countries are United States, China, Brazil and European Union, being Brazil and United States are also the main exporter countries. These two countries together provide two-thirds of global trade (FAO, 2010; FAO, 2012; USDA, 2012).

Feedstuff is an aspect of high economic importance in the rearing of commercial poultry not only because it is primarily responsible for the growth response of birds, but mainly because it represents the largest cost in the production cycle (Ávila et al., 1992). For instance, the broilers' energy requirements are responsible for 70% of the cost of the ration (Skinner et al., 1992) and, besides, the processing method and the grain type interfere differently on the economic viability and animal performance. The advantages of using processed feed have been well documented, although they represent a high cost for manufacturing. Under natural conditions, birds have to deal with different types of feed, which have different energy and protein levels. Despite domestication and selection for fast growth, broiler chickens did not lose their ability to discriminate different types of diets (Emmans & Kyriazakis, 2001). It has been suggested that the birds associate the feed physical characteristics with nutritional content, which indicates that the contact perception contributes to the identification of the feed.

Most researches on performance and behaviour of broiler chicken feeding has been with respect to productivity indices and physiological responses, but there is a lack of scientific knowledge of the biomechanical features of the bird feeding process. Chickens present cranial kinesis, which is characterized by the movement of the upper jaw in relation to the skull, a key factor in feeding efficiency found in all species of birds (Bock 1964; Zweers 1982; Feduccia 1986; Bout & Zweers, 2001; Gurd 2006; Estrella & Masero, 2007; Gurd, 2007). Past and recent publications have reported this feature in many species of fish, rodents and birds, as well as in humans. In domestic chickens (*Gallus gallus domesticus*); however, even though a few studies are found regarding the biomechanical issues of the intake process, no one is related to the modern captive breed strains for egg or meat production.

This review paper aims to approach what is known to date about the biomechanical features of the feeding behaviour of chickens. It addresses issues related to feed characteristics used in commercial broiler chicken production, with regard to feed particle size, physical form and the influence of feeders.

General Concepts of Biomechanics and Historical Context

Biomechanics can be defined as the study of the mechanical model of the body and its movements, integrating physics and biology (Domenici & Blake, 2000), or as the mechanics of movement in living creatures, being a discipline of biology that combines biophysics, physiology, physics, engineering and medicine (Low & Reed, 1996), or even simple physical (mechanical) movement displayed or produced by biological systems (Mclester & Pierre, 2008). Despite biomechanics being a relatively young discipline recognized in scientific research, its considerations are also of interest to several other scientific disciplines and professional fields, such as zoology, medical (orthopaedics, cardiology, sports medicine, physiotherapy), biomedical engineering or biomechanics, or kinesiology (study of human movement) (Hall, 1999).

Giovanni Borelli (1608-1679) is considered a pioneer in the studies of biomechanics. He integrated physiology and physical science to describe the human and animal movements, and offered thoughts on the function of muscles. The invention of the light microscope in the latter part of the seventeenth century greatly aided the study of physiology, but the advent of photography in the nineteenth century played a key role, and allowed a more detailed study of human and animal locomotion. Some knowledge of electricity was also developed in this period, which led to the use of electrical stimulation and electromyography. In the twentieth century, the invention of the electron microscope influenced the understanding of mechanical changes on a cellular level (Low & Reed, 1996).

Currently, biomechanics is seen as an academic subject and with the advancement of computer and microelectronics it is now possible to use measurement systems in more complex fields. High resolution cameras, high storage capacity and digital image processing for a relatively affordable cost make the transformation of qualitative for quantitative techniques possible, with a level of accuracy comparable to traditional punctual measuring methods. In this sense, the high speed camera is an apparatus that has been highlighted for its effectiveness in several areas of study, including animal behaviour assessments.

The Study of Biomechanics and Motion Analysis

In the study of biomechanics, it must be consider the consequences of movements produced by forces, integrating biological features with traditional mechanics (the effect of forces and energy in the motion of bodies). The static and the dynamic are two sub-branches of mechanics used to study the anatomical and functional aspects of living organisms. Static is the study of systems that are in a state of constant motion, i.e.both at rest (without movement) or in motion at a constant speed. Dynamics is the study of systems in which the acceleration is present. Kinematics and kinetics are subdivisions of biomechanical study. Kinematics is the description of motion features including the pattern and velocity of the body segments which generally translates the degree of coordination that an individual displays. Whereas kinematics describes the appearance of movement, kinetics is the study of forces associated with movement (Hall, 1999; Serway & Jewett, 2004). Anthropometric factors, e.g. size, shape and weight of body segments, are other important concerns in kinetic analysis (Hall, 1999).

Among other essential purposes, animals depend mainly on muscles to propel themselves for locomotion and food handling. Muscles are biological motors that consume chemical energy and perform mechanical work. Generally the function of muscles is considered within the 'metabolism' together with other processes, e.g.thermoregulation, which also consumes oxygen and generates heat. The power of muscles is generally viewed only by the capacity of enzyme energy supply. However, the rate at which muscles can perform the work is limited by three variables: the stress it may exercise, the tension and the contraction frequency. These are the mechanical variables, and their maximum values are defined by mechanical limitations (Pennycuick, 1992).

Nowadays biomechanics can be considered a "tool" to investigate matters of ecology, physiology and evolution. It also can be useful for assessments, forecasts and understanding of behaviours. Some structures of animals (e.g. jaw, teeth, claws, beaks and horns) may be regarded as tools and/or weapons with certain physical characteristics, and the kind of forces applied may influence their utilization. These forces can be used to handle, break or tear the food; for different ways of feeding (suction, crushing and handling through the jaw); for biting, cutting the skin, breaking bones or killing (Domenici & Blake, 2000). Several factors affect the execution of eating action, such as competition, energy consumption, risk of predation, prey availability and predator performance. Performance includes the ability of a predator to locate, capture and manipulate the prey, all being influenced by their morphology (Wainwright, 1991).

Biomechanical studies have been widely investigated using high speed camera technology in various species of animals, e.g. insects (Dangles et al., 2006; Wu et al., 2008; Nguyen et al., 2010; Truong et al., 2012); fish (Korff & Wainwright, 2004; Herrel et al., 2005; Huber et al., 2008; Wroe et al., 2008; Huber et al., 2009; Maraet al., 2009; Habegger et al., 2010; Tran et al., 2010); rodents (Bracha et al., 2003; Sakatani & Isa, 2004; Herbin et al., 2007; Morita et al., 2008; Beare et al., 2009; Fu et al., 2009; Stefen et al., 2011); reptiles (Deban & O'Reilly, 2005; Herrel & O'Reilly, 2006; Fuller et al., 2011;

Schaerlaeken et al., 2011); birds (Westneat et al., 1993; Estrella & Masero, 2007; Abourachid et al., 2011; Dawson et al., 2011; Smith et al., 2011); as well as in humans (Arampatzis et al., 1999; Yoganandan et al., 2002; Imura et al., 2008; Shan, 2008; Bakker et al., 2009; Steeve, 2010). The main topics treated are flight features, bite force analysis, cognitive functions assessments by real-time tracking, anatomical and physiological study of locomotion, evaluation of mandibular motion and muscle activity during ingestion or vocalization, the effect of food type on feeding efficiency, 3-D bones reconstruction for motion morphology assessments, among others.

At some time, several reasons induced the domestication of birds. These include: communication (pigeon); vestment (ostrich), sport (falcon), decoration (peafowl), religion (Egyptian goose); and pet (cage birds). Nowadays, the main aims of domestication are egg and meat production. Economically, these activities are very important, since producing poultry meat, and eggs are very efficient ways to transforming vegetable mass into meat protein (FAO, 2010). In the upper limbs, the birds have wings moved by powerful pectoral muscles, consisting of a very well developed structure and the skeletal bones are significantly lighter. These features have given the birds a high mobility, allowing their dispersion throughout the environment and consequently their adaptation to a variety of environments. These adjustments led to different types of secondary anatomic variations of the beak, oral cavity, feathers, wings, legs and feet (King, 1986). Thus, a better comprehension of the biomechanics of each element is helpful for studying disease aetiology, and for making treatment decisions and general motion assessments.

On the other hand, some methodological drawbacks could be encountered when it is necessary to adopt a surgical intervention for implant insertion, which could involve ethical concerns and technological limitations (Bergmann et al., 2001; Stansfield et al., 2003), beyond the stress to which the individual could be subjected. In addition, the labour intensiveness, utilization of electrical stimulation and post mortem examination can lead to a non-real situation, such as the lack of functional movements (Gussekloo et al., 2001). Developing a precise and non-invasive method for measuring the internal force within the living body still remains a great challenge in the field of biomechanics and motion analysis (Lu & Chang, 2012). Motion analysis can be an effective method for identifying beneficial and damaging elements when a moving system of a living organism is performing a task. Some advantage via the utilization of high speed cameras and computational image analysis for motion assessments has been achieved, especially with respect to its relatively low cost, versatility in analysis, commercial availability of the hardware and possibility of system upgrade according to need (Sakatani & Isa, 2004).

Chicken Intake Process: Anatomical and Biomechanical Approaches

The digestive system of the chicken is considered simple, short and extremely efficient. The beak collects the food, and the bird decides whether to accept or reject it through the tactile cells. This decision is based on reflectivity and taste, even though the number of taste buds is small. No evidence has been produced to suggest that chickens have any real ability to smell. The food is swallowed whole with a little saliva, through the oesophagus to the crop, in which the fibre is softened, and the food is acidified by lactic acid. From the crop, the food passes into the proventriculus, which secretes acid and pepsin, an organ that best resembles the stomach of a mammal. Thereafter the food passes into the gizzard, an organ with powerful muscles that contract rhythmically to reduce the thickness of the content. After that, the food passes through various regions of the intestine by peristaltic contractions, and it is at this stage that digestion and nutrient absorption occur. The digestion also occurs to a lesser extent in the caeca, two bags that are located at the junctions of the small and large intestines, the latter being responsible for the absorption of water. From here the faeces move into the cloaca for evacuation, which is also related to the excretion of urine, acceptance of delivery of sperm and the passage of egg outwards (Sainsbury, 1980).

The birds have one of the most skilled skulls of living vertebrates, besides the pneumatisation by epithelial extensions of air sacs, a fact that allows alleviates the weight, they are kinetic. The cranial kinesis is related to the movement of the upper jaw, or part of it, in relation to the skull, which is a characteristic found in all species of birds (Bock, 1964; Zweers, 1982; Feduccia, 1986; Bout & Zweers, 2001; Gussekloo & Bout, 2005). This is not an exclusive feature of birds, as it is also found in fish, reptiles and amphibian fossils (Bock, 1964). The skull of birds can be divided into functional units: the braincase, the upper jaw, the bone structure that comprises the palate, the jugal bar and quadrate, and the lower jaw. These functional units operate together in which the quadrate bone plays a key role during the beak movement (Van Den Heuvel, 1992).

There are many proposed functions of cranial kinesis which can be highlighted: the highest elevation of the upper jaw, reducing the force required to open the beak, keeping the beak closed without muscular effort, higher beak closing speed, shock absorption, increased capacity of food selection, maintenance of the primary axis of orientation and attachment of the buccal apparatus muscles (Bock, 1964; Bout & Zweers, 2001; Gurd, 2006; Estrella & Masero, 2007; Gurd, 2007). Furthermore, the cranial kinesis can be uncoupled or coupled. Uncoupled is when the upper and lower jaws move independently. Coupled kinetics occurs owing to two separate mechanisms, or a combination of both. In most birds, the presence of postorbital ligaments and the lacrymomandibular is the main morphological feature of this system. When one of these ligaments is stretched to the maximum, the lower jaw cannot be depressed without the quadrate bone swinging forward while the opposite occurs in beak closing motion, establishing a relationship of dependence of both upper and lower jaws, although a certain degree of independence in this mechanism may exist (Bock, 1964).

The domestic fowl has a prokinetic skull mainly characterized by a postorbital ligament, also known as the squamosomandibular ligament, whereby the skull connects with the mandibular process. Other species can also present rhincokinetic or amphikinetic skulls, differing in the location of the jaw joint. Therefore, the chicken jaw is a unique structure that moves entirely. When the beak is usually closed, the ligaments are not tensioned, and the system is considered at rest. The coupled cranial kinesis in domestic fowl does not play a dominant role in the feeding process. The jaw is lowered 20ms after the lifting of the upper jaw, indicating that the coupled cranial kinesis does not occur while the food is grasped, but can occur eventually. Similar characteristics may occur in subsequent cycles for the transport of food into the oral cavity during the food manipulation. However, the coupled kinesis is used when the bird closes its beak, as it is not possible to depress the upper jaw without raising the lower jaw (Van Den Heuvel, 1992).

The feeding behaviour of animals can be divided into appetitive phases, corresponding to the demand for feed and consummatory act, which is the real feed intake. The assessments may be related to bite events and/or visits to feeders (Slater, 1974; Berdoy, 1993; Nielsen, 1999) in which these could be considered as a unit to analyze feeding behaviours (Nielsen & Whittemore, 1995). There is no real chewing in birds, the tongue is rigid and tactile sensibility is mainly perceived when the particles are touched and seized by the beak tip (Picard et al., 2002). The appetitive phase of chickens can be characterized by the foraging behaviour, which is the time that the bird explores the environment searching for food, as reported by (Yo et al., 1997), who found that two thirds of young bird pecks do not result in the prehension of a feed particle.

The mechanical process of feeding in domestic chicken is similar to that of pigeons (Table 1).

Table 1: Summarized description of the phases of the pigeon and chicken feeding scenario.

Phase	Description
Fixation	The head still stable above the seed, the eyes wide open. The distance between the eye and the target is about 5-8 cm. Beak is closed, but fluctuations could be seen in the openings of the beak and tongue movements for swallowing seeds ingested previously.
Approach or pecking	Starts when the bird moves its head uninterruptedly towards food in an oblique or vertical direction. The beak opens, and the elevation of the upper jaw occurs prior to the depression of the lower jaw and the tongue is retracted. The beak opens slightly more than the seed size. The eyes are partially closed.
Grasping	Starts with the maximum beak opening in the last part of the approach phase. The beak tip apprehends the seed and the eyes are completely closed.
Withdrawal	Starts right after the 'grasping' phase. Food is retained in the beak tip, and head is withdrawn in an upward motion. There may be a delay when the beak strikes against the substrate.
Stationing	The food is eventually repositioned by "catch-and-throw" movements. These serve to reposition the seed in the beak before starting the transport. This phase can be repeated as often as needed or possibly skipped when seed is properly grasped.

Transporting	Transports the seed from the beak tip into the pharynx level though the "catch-and-throw" or "slide-and-glue" movements or a combination of both. The "slide-and-glue" technique, usually adopted with smaller particles, consists in the displacement of the tongue up to the tip of the beak in order to glue the food with the aid of the sticky saliva and convey it into the oral cavity.
Collecting	Small seeds are accommodated at the base of the tongue while the bird keeps feeding. It does not occur with large seeds
Swallowing	Final transportation of the seed into the esophagus with one or more movements of the pharynx, tongue, small beak openings and head jerks. Two mechanisms: "scraping" which is the continuation of "slide-and-glue" for small seeds and "peristaltic" which is a continuation of "catch-and-throw" for larger seeds.

Adapted from (Moon & Zeigler, 1979; Zeigler et al., 1980; Zweers, 1982; Bermejo et al., 1989; Van Der Heuvel & Berkhoudt, 1998).

It is suggested that, within the phases 'grasp' and 'mandibular motion', the opening beak amplitude is gauged according to the particle size and the initial beak opening is used to control the amplitude. For the 'grasp' phase, the birds use visual information and for 'mandibular motion' tactile information. Moreover, the feeding behaviour of these birds can be defined as stereotyped movement patterns. These stereotyped patterns create an eating-response sequence and such sequences create an event feeding scene or a feeding bout (Figure 1). The reason why these movements are defined as stereotyped is on account both of duration and temporal organization of the variables in the process. This standard is based on the Variation Coefficient. Considering the appearance of stereotyped variables that compose a feeding scene of pigeons (Zweers, 1982) and chickens (Van Der Heuvel & Berkhoudt, 1998), the feeding behaviour can be considered as a result of Fixed Action Patterns, more than just a pattern. Actually, the bird can adapt certain movement patterns depending on the type of food, but such behaviours are subordinate to limitations of morphological structure and mechanical construction (Zweers, 1982).

REFERENCES

1. Aarseth KA, Prestløkken E. Mechanical properties of feed pallets: Weibull analysis. Biosystems Engineering 2003; 3: 349-361

2. Abourachid A, Hackert R, Herbin M, Libourel PA. Lambert, F.; Gioanni, H.; Provini, P.; Blazevic, P.; Hugel, V. Bird terrestrial locomotion as revealed by 3D kinematics. Zoology 2011; 114: 360-368

3. Addo A, Bart-Plange A, Akowuah JO. Particle size evaluation of feed ingredient produced in the Kumasi Metropolis, Ghana. ARPN Journal of Agricultural and Biological Science 2012; 7(3): 177-181

4. Agroceres, 2004. Manual de Manejo de Frangos Agrosseres, s.l.: AGROSSERES ROSS

5. Albino JJ, Bassi L, Saatkamp M. 2011. Regulagem e distribuição de comedouros tubulares e bebedouros pendulares em aviários convencionais. [Online] Available at: ALBINO, J. J.; BASSI, L.; SAATKAMP, M. Regulagem e distribuição de comedouros tubulares e bebedouros pendulares em aviários convencionais. Erhttp://pt.engormix.com/MA-avicultura/ad [Acessed: 14 August 2012

6. Amerah A M, Ravindran V, Lentle RG, Thomas DG. Feed particle size: Implications on the digestion and performance of poultry. World's Poultry Science Journal 2007; 63: 439-451

7. Amerah AM, Ravindran V, Lentle RG, Thomas DG. Influence of feed particle size on the performance, energy utilization, digestive tract development, and digesta parameters of broiler starters fed wheat- and corn-based diets. Poultry Science 2008; 87: 2320-2328

8. Angulo E, Brufau J, Esteve-Garcia E. Effect of a sepiolite product on pellet durability in pig diets differing in particle size and in broiler starter and finisher diets. Animal Feed Science Technology 1996; 63: 25-34

9. Arampatzis A, Brüggemann G, Metzler V. The effect of speed on leg stiffness and joint kinetics in human running. Journal of Biomechanics 1999; 32: 1349-1353.

10. Arce-Menocal J, Avila-Gonzales E, Lopez-Coelho C, Garibay-Torres L, Martinez-Lemus LA. Body weight, feed-particle size, and ascites incidence revisited. The Journal of Applied Poultry Research 2009; 18: 465-471]

11. ASAE - American Society of Agricultural Engineers, 1983. Wafers, pellets, crumbles-definitions and methods for determining density, durability, and moisture content, St. Joseph: American Society of Agricultural Engineers

12. ASAE - American Society of Agricultural Engineers, 2003a. Cubes, pellets, and crumbles – definitions and methods for determining density, durability and moisture content, St. Joseph: ASABE

13. ASAE - American Society of Agricultural Engineers, 2003b. Method of determining and expressing fineness of feed materials by sieving, St. Joseph: ASABE

14. Aviagen, 2009. Ross Broiler Management Manual, s.l.: s.n

15. Ávila VS, Jaenisch FRF, Pieniz LC, Ledur MC, Albino LFT, Oliveira PAV. Produção e manejo de frangos de corte, Concórdia: Embrapa Suínos e Aves, 1992

16. Ávila VS, Kunz A, Bellaver CP, Paiva D, Jaenisch FRF, Mazzuco H, Trevisol IM, Palhares JCP, Abreu PG, Rosa PS. Boas Práticas de Produção de Frangos de Corte. Concórdia: Embrapa Suínos e Aves. 2006

17. Axe DE. Factors affecting uniformity of a mix. Animal Feed Science and Technology 1995; 53: 211-220

18. Bakker PP, Manske SL, Ebacher V, Oxland TR. Cripton PA, Guy P. During sideways falls proximal femur fractures initiate in the superolateral cortex: Evidence from high-speed video of simulated fractures. Journal of Biomechanics 2009; 42: 1917-1925

19. Bassi LJ, Albino JJ, Ávila VS, Shmidt GS, Jaenisch FRF. Recomendações Básicas para Manejo de Frangos de Corte Colonial, Concórdia: s.n., 2006

20. Beare JE, Morehouse JR, DeVries WH, Enzmann GU, Burke DA, Mag-nuson DSK, Whittemore SR. Gait analysis in normal and spinal contused mice using the treadscan system. Journal of Neurotrauma 2009; 26: 20452056.

21. Behnke KC. Feed manufacturing technology: current issues and challenges. Animal Feed Science and Technology 1996; 62(1): 49-57.

22. Behnke KC. Factors influencing pellet quality. Feed Technology 2001; 5(4): 9-22

23. Berdoy M. Defining bouts of behaviour: a three process model. Animal Behaviour 1993; 46: 387-396

24. Bergmann G, Deuretzbacher G, Heller M, Graichen F, Rohlmann A, Strauss J, Duda GN. Hip contact forces and gait patterns from routine activities. Journal of Biomechanics 2001; 34: 859-871

25. Bermejo R, Allan RW, Houben D, Deich JD, Zeigler HP. Prehension in the pigeon I: descriptive analysis. Experimental Brain Research 1989; 75: 569–576

26. Bock WJ.. Kinetics of the avian skull. Journal of Morphology 1964; 114: 1-42

27. Bout RG, Zweers GA. The role of cranial kinesis in birds. Comparative Biochemistry and Physiology 2001; 131A: 197-205

28. Bracha V, Nilaweera W, Zenitsky G, Irwin K. Video recording system for the measurement of eyelid movement during classical conditioning of the eyeblink response in the rabbit. Journal of Neuroscience Methods 2003; 125: 173-181

29. Brickett KE, Dahiya JP, Classen HL, Annett CB, Gomis S. The Impact of Nutrient Density, Feed Form, and Photoperiod on the Walking Ability and Skeletal Quality of Broiler Chickens. Poultry Science 2007; 86: 2117-2125

30. Briggs JL, Maier DE, Watkins BA, Behnke KC. Effect of ingredients and processing parameters on pellet quality. Poultry Science 1999; 78: 1464–1471

31. Buskirk DD, Zanella AJ, Harrigan TM, Van Lente JL, Gnagey LM, Kaercher MJ. Large round bale design affects hay utilization and beef cow behavior. Journal of Animal Science 2003; 81: 109-115

32. Carré B, Muley N, Gomez J, Oury F-X, Laffitte E, Guillou D, Signoret C. Soft wheat instead of hard wheat in pelleted diets results in high starch digestibility in broiler chickens. British Poultry Science 2005; 46 (1): 66–74

33. Clark PM, Behnke KC, Fahrenholz AC. Effects of feeding cracked corn and concentrate protein pellets on broiler growth

performance. The Journal of Applied Poultry Research 2009; 18: 259-268

34. Cobb-Vantress. COBB Broiler Management Guide, s.l.: COBB-VANTRESS, 2010

35. Corzo A, Mejia L, Loar II RE. Effect of pellet quality on various broiler production parameters. The Journal of Applied Poultry Research 2011; 20: 68-74

36. Cutlip SE, Hott JM, Buchanan NP, Rack AL, Latshaw JD, Moritz JS. The Effect of Steam-Conditioning Practices on Pellet Quality and Growing Broiler Nutritional Value. The Journal of World›s Poultry Research 2008; 17: 249–261

37. Dahlke F, Ribeiro AML, Kessler AM, Lima ARR, Maiorka A. Effect of corn particle and physical form of the diet on the gastrointestinal structures of broiler chickens. Brazilian Journal of Poultry Science 2003; 5: 61-67

38. Dahlke F, Ribeiro AML, Kessler AM, Lima AR. Corn particle size and physical form of the diet and their effects on performance and carcass yield of broilers. Brazilian Journal of Poultry Science 2001; 4(3): 241-248

39. Dangles O, Ory N, Steinmann T, Christides JP, Casas J. Spider›s attack versus cricket›s escape: velocity modes determine success. Animal Behaviour 2006; 72: 603-610

40. Dawson MM, Metzger KA, Baier DB, Brainerd EL. Kinematics of quadrant bone during feeding in mallard ducks. The Journal of Experimental Biology 2011; 214: 2036-2046

41. Deban SM, O›Reilly JC. The ontogeny of feeding kinematics in a giant sala-mander ryptobranchus alleganiensis: Does current function or phylogenetic relatedness predict the scaling patterns of movement?. Zoology 2005; 108: 155-167

42. Domenici P, Blake RW. Biomechanics in animal behaviour. 1 ed. Oxford: BIOS Scientific Publishers Ltd

43. Donald AM. Plasticization and self-assembly in the starch granule. Cereal Chemistry 2001; 78: 307-314

44. Douglas JH, Sullivan TW, Bond PL, Struwe FJ, Baier JG, Robeson LG. Influence of grinding, rolling, and pelleting on the nutritional value of grain sorghums and yellow corn for broilers. Poultry Science 1990; 69: 2150-2156

45. Dozier III W. Cost-effective pellet quality for meat birds. Feed Management, February, 2001; 52(2), p. 3

46. Dozier III WA, Behnke KC, Gehring CK, Branton SL. Effects of feed form on growth performance and processing yields of broiler chickens during a 42-day production period. The Journal of Applied Poultry Research 2010; 19: 219-226

47. Dozier III WA, Behnke K, Twining P, Branton SL. Effects of the addition of roller mill ground corn to pelleted feed during a fifty-six-day production period on growth performance and processing yields of broiler chickens. The Journal of Applied Poultry Research 2009; 18: 310-317

48. Emmans G, Kyriazakis I. Consequences of gentic change in farm animals on food intake and feeding behaviour. 2001; s.l., s.n., 115-125

49. Engberg RM, Hedemann MS, Jensen BB. The influence of grinding and pelleting of feed on the microbial composition and activity in the digestive tract of broiler chickens. British Poultry Science 2002; 44: 569-579

50. Englert S. Avicultura: Tudo sobre raças, manejo e alimentação. 7 ed. Guaíba (Rio Grande do Sul): Editora Agropecuária, 1998

51. Estrella SM, Masero JA. The use of distal rhyncokinesis by birds feeding in water. The Journal of Experimental Biology 2007; 210: 3757-3762

52. Fairfield DA. Pelleting for Profit - Part 1. Feed and Feeding Digest 2003; 54(6): 1-5

53. FAO, 2010. Poultry meet & eggs., Rome: FAO

54. FAO, 2012. Food Outlook, Rome: Trade and Market Division of FAO

55. Faria DE, Faria Filho DE, Junqueira OM, Araújo LF, Torres KAA. Forma física e níveis de energia metabolizável a ração para frangos de corte de 1 a 21 dias de idade. ARS VETERINARIA - Revista de Medicina Veterinaria e Zootecnia 2006; 22(3): 259-264

56. Feduccia A. 1986. Osteologia das aves. In: Anatomia dos animais domésticos. 5 ed. Rio de Janeiro: Guanabara Koogan, 1680-1689; 1986

57. Ferket PR, Gernat AG. Factors that affect feed intake of meat birds: A Review. International Journal of Poultry Science 2006; 5(10): 905-911

58. Freire R, Walker A, Nicol CJ. The relationship between trough height, feather cover and behaviour of laying hens in modified cages. Applied Animal Behaviour Science 1999; 63: 55-64

59. Freitas ER, Sakomura NK, Dahlke F, Santos FR, Barbosa NAA. Desempenho, eficiência de utilização dos nutrientes e estrutura do trato digestório de pintos de corte alimentados na fase pré-inicial com rações de diferentes formas físicas. Revista Brasileira de Zootecnia 2008; 37: 73-78

60. Freitas ER, Sakomura NK, Vieira RO, Neme R, Traldi AB. Uso de diferentes formas físicas e quantidades de ração pré-inicial para frangos de corte. Revista Ciência Agronômica 2009; 40(2): 293-300

61. Fukuoka M, Ohta KI, Watanabe H. Determination of the terminal extent of starch gelatinization in a limited water system by DSC. Journal of Food Engineering 2002; 53: 39-42

62. Fuller PO, Higham TE, Clark, AJ. Posture, speed, and habitat structure: three-dimensional hindlimb kinematics of two species of padless geckos. Zoology 2011; 114: 104-112

63. Fu SC, Chan KM, Chan LS, Fong, DTP, Lui PYP. The use of motion analysis to measure pain-related behaviour in a rat model of degenerative tendon injuries. Journal of Neuroscience Methods 2009; 179: 309-318

64. Gabriel I, Mallet S, Leconte M, Fort G, Naciri M. Effects of whole wheat feeding on the development of coccidial infection in broiler chickens until market-age. Animal Feed Science and Technology 2006; 129: 279-303

65. Gadzirayi CT, Mutandwa E, Chihiya J, Mlambo R. A comparative economic analysis of mash and pelleted feed in broiler production under deep litter housing system. International Journal of Poultry Science 2006; 7: 629-631

66. Garcia Neto M, Campos EJ. Incidência de ascite em frangos de corte alimentados com rações comerciais de alto nível energético. Pesquisa Agropecuária Brasileira 2002; 37: 1205-1212

67. GLOBOAVES, 2011. Manual de Manejo Linha Colonial, s.l.: s.n

68. Goodband RD, Tokach MD, Nelssen JL. The effects of diet particle size on animal performance., Kansas: MF-2050 Feed Manufacturing. 2002

69. Greenwood MW, Cramer KR. Clark PM, Behnke KC, Beyer RS. Influence of feed form on dietary lysine and energy intake and utilization of broilers from 14 to 30 days of age. International Journal of Poultry Sciences 2004; 3: 189-194

70. Gurd DB. Filter-feeding dabbling ducks (*Anas* s) can actively select particles by size. Zoology 2006; 109: 120-126

71. Gurd DB. Predicting resource partitioning and community organization of filter-feeding dabbling ducks from functional morphology. The American Naturalist 2007; 169: 334-343

72. Gussekloo SWS, Bout RG. The kinematics of feeding and drinking in paleognathous birds in relation to cranial morphology. Journal of Experimental Biology 2005, 208: 3395-3407

73. Gussekloo SWS, Vosselman MG, Bout RG. Three-dimensional kinematics of skeletal elements in avian prokinetic and rhynchokinetic skulls determined by Roentgen stereo-photogrammetry. Journal of Experimental Biology 2001, 204: 1735-1744

74. Habegger ML, Motta PJ, Huber DR, Deban SM.. Feeding biomechanics in the Great Barracuda during ontogeny. Journal of Zoology 2010; 283: 63-72

75. Hall SJ. Basic Biomechanics. 3ª ed. Singapore: Mcgraw-Hill; 1999

76. Hamilton RMG, Proudfoot FG. Ingredient particle size and feed texture effects on the performance of broiler chickens. Animal Feed Science and Technology 1995; 51: 203-210

77. Herbin M, Hackert R, Gasc JP, Renous S. Gait parameters of treadmill versus overground locomotion in mouse. Behavioural Brain Research Journal 2007; 181: 173–179.

78. Herrel A, O'Reilly JC. Ontogenetic scaling of bite force in lizards and turtles. Physiological and Biochemical Zoology 2006; 79: 31–42

79. Herrel A, Van Wassenbergh S, Wouters S, Aerts P, Adriaens D. A functional morphological aroach to the scaling of the feeding

system in the African catfish, *Clarias gariepinus*. The Journal of Experimental Biology 2005; 208: 2091-2102

80.   Hetland H, Choct M, Svihus B. Role of insoluble non-starch polysaccharides in poultry nutrition. World's Poultry Science Journal 2004; 60: 415-422

81.   Hetland H, Svihus B. Effect of oat hulls on performance, gut capacity and feed passage time in broiler chickens. British Poultry Science 2001; 42: 345-361

82.   Hetland H, Svihus B, Olaisen V. Effect of feeding whole cereals on performance, starch digestibility and duodenal particle size distribution in broiler chickens. British Poultry Science 2002; 43(3): 416-423

83.   Huber DR, Claes JM, Mallefet J, Herrel A. Is extreme bite performance associated with extreme morphologies in sharks?. Physiological and Biochemical Zoology 2009; 82: 20-28

84.   Huber DR, Dean MN, Summers AP. Hard prey, soft jaws and the ontogeny of feeding mechanics in the spotted ratfish *Hydrolagus colliei*. Journal of the Royal Society Interface 2008; 5: 1-12

85.   Imura A, Iino Y, Kojima T. Biomechanics of the continuity and speed change during one revolution of the Fouette turn. Human Movement Science 2008; 27: 903-913

86.   Ipek A, Sahan U, Yilmaz B. The effect of drinker type and drinker height on the performance of broiler cockerels. Czech Journal of Animal Science 2002; 47(11) : 460-466

87.   Jensen LS. Influence of pelleting on the nutritional needs of poultry. Asian-Australasian Journal of Animal Science 2000; 13: 35-46

88.   Jones F.T, Anderson KE, Ferket PR. Effect of extrusion on feed characteristics and broiler chicken performance. The Journal of Applied Poultry Research 1995; 4: 300-309

89.   Jones GPD, Taylor RD. The incorporation of whole grain into pelleted broiler chicken diets: Production and physiological responses. British Poultry Science 2001; 42: 477-483

90.   Joy MT, DePeters EJ, Gadel JG, Zinn RA. Effect of corn processing on the site and extent of digestion in lactating cows. Journal of Dairy Science 1997; 80: 2087-2097

91. King AS. Introdução às aves. In: Anatomia dos animais domésticos. 5 ed. Rio de Janeiro: Guanabara Koogan, 1677-1679; 1986

92. Kishida T, Nogami H, Himeno S, Ebihara K. Heat moisture treatment of high amylose corn starch increases its resistant starch content but not its physiological effects in rats. Journal of Nutrition 2001; 131: 27162721

93. Koch K. Hammermills and rollermills, Kansas: MF-2048 Feed Manufacturing, 1996

94. Korff WL, Wainwright PC. Motor pattern control for increasing crushing force in the striped burrfish (*Chilomycterus schoepfi*). Zoology 2004; 107: 335–346

95. Lara LJC, Baião NC, Rocha JSR, Lana AMQ, Cançado SV, Fontes DO, Leite RS. Influência da forma física da ração e da linhagem sobre o desempenho e rendimento decortes de frangos de corte. Arquivo Brasileiro de Medicina Veterinária 2008; 60(4): 970-978

96. Lecznieski JL, Ribeiro AML, Kessler AM, Penz Jr, AM. Influence of physical form and energy level of the diet on performance and carcass composition of broilers. Archivos Latinoamericanos de Produccion Animal 2001; 9(1): 6-11

97. Leeson S, Caston LJ, Summers JD, Lee KH.. Performance of male broilers to 70 days when feed diets of varying nutrient density as mash or pellets. The Journal of Applied Poultry Research 1999; 8: 452-464

98. Lemme A, Wijtten PJA, Van Wichen J, Petri A, Langhout DJ. Responses of male growing broilers to increasing levels of balanced protein offered as coarse or pellets of varying quality. Poultry Science 2006; 85(04): 721-730

99. Lentle RG, Ravindran V, Ravindran G, Thomas DV. Influence of feed particle size on the efficiency of broiler chickens fed wheat based diets. Journal of Poultry Science 2006; 43: 135-142

100. Lilburn M.S. Pratical aspects of early nutrition for poultry. The Journal of Applied Poultry Research 1998; 7: 420-424

101. Lilly KGS, Gehring CK, Beaman KR, Turk PJ, Sperow M, Moritz JS. Examining the relationships between pellet quality, broiler performance, and bird sex. The Journal of Applied Poultry Research 2011; 20: 231–239

102. López CAA, Baião NC. Efeitos do tamanho da partícula e da forma física da ração sobre o desempenho, rendimento de carcaça e peso dos órgãos digestivos de frangos de corte. Arquivo Brasileiro de Medicina Veterinária 2004; 56(2): 214-221

103. López CAA, Baião NC, Lara LJC, Rodriguez NM, Cançado SV. Efeitos da forma física da ração sobre a digestibilidade dos nutrientes e desempenho de frangos de corte. Arquivo Brasileiro de Medicina Veterinaria e Zootecnia 2007; 59(4): 1006-1013

104. Lott BD, May JD, Simmons JD, Branton SL. The effect of nile height on broiler performance. Poultry Science 2001; 80: 408-410

105. Low J, Reed A. Basic biomechanics explained. Oxford: Butterworth-Heinemann Ltd.; 1996

106. Lu T, Chang C. Biomechanics of human movement and its clinical alications. Kaohsiung Journal of Medical Sciences 2012; 28: S13-S25

107. Maiorka A, Dahlke F, Penz AM, Kessler AM.. Diets formulated on total or digestible amino acid basis with different energy levels and physical form on broiler performance. Brazilian Journal of Poultry Science 2005; 7(1): 7-50

108. Mara KR, Motta PJ, Huber DR. Bite force and performance in the durophagous bonnethead shark, Sphyrna tiburo. Journal of Experimental Zoology 2009; 311A:1-11

109. May JD, Lott BD, Simmons JD. Water consumption by broilers in high cyclic temperatures: bell versus nile waterers. Poultry Science 1997; 76: 944 947

110. McKinney L, Teeter R. Predicting effective caloric value of nonnutritive factors: I. pellet quality and II. prediction of consequential formulation dead zones. Poultry Science 2004; 83: 1165-1174

111. Mclester J, Pierre PS. Applied Biomechanics: concepts and connections. Belmont: Thomson Wadsworth; 2008

112. Meinerz C, Ribeiro AML, Penz Jr. AM, Kessler AM. Níveis de Energia e peletização no desempenho e rendimento de carcaça de frangos de corte com oferta alimentar equalizada. Revista Brasileira de Zootecnia 2001; 30 (6S): 2026-2032

113. Meurer RP, Fávero A, Dahlke F, Maiorka A. Avaliação de rações peletizadas para frangos de corte. Archives of Veterinary Science 2008; 13(3): 229-240

114. Moon RD, Zeigler HP. Food preferences in the pigeon (*Columba livia*). Physiology & Behavior 1979; 22(6): 1171–1182

115. Moreira I, Rostagno HS, Coelho DT, Costa PMA, Tafuri ML. Determinação dos coeficientes de digestibilidade, valores energéticos e índices de controle de qualidade do milho e soja integral processados a calor. Revista Brasileira de Zootecnia 1994; 23: 916-929

116. Morita T, Fujiwara T, Negoro T, Kurata C, Maruo H, Kurita K, Goto S, Hirab K. Movement of the mandibular condyle and activity of the masseter and lateral pterygoid muscles during masticatory-like jaw movements induced by electrical stimulation of the cortical masticatory area of rabbits. Archives of Oral Biology 2008; 53: 462-477

117. Moritz JS, Beyer RS, Wilson KJ, Cramer KR, McKinney LJ, Fairchild FJ. Effect of moisture addition at the mixer to a corn-soybean-based diet on broiler performance. The Journal of Applied Poultry Research 2001; 10: 347-353

118. Neves DP, Nääs IA, Vercellino RA, Moura DJ. Do broilers prefer to eat from a certain type of feeder?. Revista Brasileira de Ciencia Avícola 2010; 12(3): 179-187

119. Nguyen QV, Park HC, Goo NS, Byun D. Characteristics of a Beetle›s Free Flight and a Flaing-Wing System that Mimics Beetle Flight. Journal of Bionic Engineering 2010; 7: 77-86

120. Nielsen BL. On the interpretation of feeding behaviour measures and the use of feeding rate as an indicator of social constraint. Applied Animal Behaviour Science 1999, 63: 79-91

121. Nielsen BLLAB, Whittemore CT. Effects of single-space feeder design of feeding behaviour and performance of growing pigs. Animal Science 1995; 1: 575-579

122. Nir I, Hillel R, Ptichi I, Shefet G. Effect of particle size on performance .3. Grinding pelleting interactions. Poultry Science 1995; 74: 771-783

123. Nir I, Hillel R, Shefet G, Nitsan Z. Effect of grain particle size on performance. 2. Grain texture interactions. Poultry Science 1994a; 73: 781-791

124. Nir I, Melcion JP, Picard, M. Effect of particle size of sorghum grains on feed intake and performance of young broilers. Poultry Science 1990; 69: 2177-2184

125. Nir I, Nitsan Z, Mahagna M. Comparative growth and development of the digestive organs and of some enzimes in broiler and egg type chicks after hatching. British Poultry Science 1993; 34: 523-532

126. Nir I, Shefet G, Aaroni Y. Effect of particle size on performance. 1. Corn. Poultry Science 1994b; 73: 45-49

127. Nir I, Twina Y, Grossman E, Nitsan Z. Quantitative effects of pelleting on performance, gastrointestinal tract and behavior of meat-type chickens. British Poultry Science 35: 589-601

128. Noy Y, Sklan D. Nutrient use in chicks during the first week posthatch. Poultry Science 1994c; 81(03): 391-399

129. Oliveira AA, Gomes AVC, Oliveira GR, Lima MF, Dias GEA, Agostino TSP, Sousa FDR, Lima CAR. Performance and carcass characteristics of broilers fed diets of different physical forms. Revista Brasileira de Zootecnia 2011; 40(11): 2450-2455

130. Parsons AS, Buchanan NP, Blemings KP, Wilson ME, Moritz JS. Effect of Corn Particle Size and Pellet Texture on Broiler Performance in the Growing Phase. The Journal of Applied Poultry Research 2006; 15: 245-255

131. Pennycuick CJ. Newton rules biology: A physical aroach to biological problems. New York: Oxford University Press, 1992

132. Perez H, Oliva-Teles A. Utilization of raw and gelatinized starch by European sea bass (Dicentrarchus labrax) juveniles. Aquaculture 2002; 205: 287-299

133. Péron A, Bastianelli D, Oury FX, Gomez J, Carre B. Effects of food deprivation and particle size of ground wheat on digestibility of food components in broilers fed on a pelleted diet. British Poultry Science 2005; 46: 223-230

134. Picard M, Melcion JP, Bertrand D, Faure JM. Visual and tactile cues perceived by chickens. London, CAB Interational, 279-298; 2002

135. PLANALTO G. Manual do Frango de Corte., s.l.: Granja Planalto, 2006

136. Plavnik I, Sklan D. Nutritional effects of expansion and short time extrusion on feeds for broilers. Animal Feed Science and Technology 1995; 55: 247-251

137. Portella FJ, Caston LJ, Leeson S. Aarent feed particle size preference by broilers. Canadian Journal of Animal Science 1988; 68: 923-930

138. Quentin M, Bouvarel I, Picard M. Short- and long-term effects of feed form on fast- and slow-growing broilers. The Journal of Applied Poultry Research 2004; 13: 540–548.

139. Quintana JÁ, Castañeda MP, Aguilera H, López C, Quiroz M, Cázares R, Ruiz R, Ávila E. Efecto de la altura de los comederos sobre el largo del tarso, pigmentación y parámetros productivos en pollo de engorda. Veterinaria México 1998; 29(1): 41-47

140. Rodgers NJ, Choct M, Hetland H, Sundby F, Svihus B. Extent and method of grinding of sorghum prior to inclusion in complete pelleted broiler chicken diets affects broiler gut development and performance. Animal Feed Science and Technology 2012; 171: 60-67

141. Roll VFB, Dai Prá MA, Roll AAP, Xavier EG, Rossi P, Anciuti MA, Rutz F. Influência da altura de comedouros tubulares no comportamento ingestivo de frangos de corte (Effect of tubular feeder height on ingestive behavior of broiler). Archivos de Zootecnia 2010a; 59 (225): 115-122

142. Roll VFB, Dai Prá MA, Xavier EG, Osório MT, Correa EK, Silveira MHD, Anciuti MA, Rutz F. Efeito da altura do comedouro tubular sobre o desempenho e qualidade de carcaça em frangos de corte no período de 28 a 42 dias de idade. Ciência Animal Brasileira 2010b; 11(4): 764-769

143. Sainsbury D. Poultry health and management. Saint Albans: Granada Publish Ltd., 1980

144. Sakatani T, Isa T. PC-based high-speed video-oculography for measuring rapid eye movements in mice. Neuroscience Research 2004; 49: 123-131

145. Schaerlaeken V, Montuelle SJ, Aerts P, Herrel A. Jaw and hyolingual movements dur-ing prey transport in varanid lizards: effects of prey type. Zoology 2011; 114: 165-170

146. Scott TA, Swift ML, Bedford MR. The influence of feed milling, enzyme sulementation, and nutrient regimen on broiler chick performance. The Journal of Applied Poultry Research 1997; 6: 391-398

147. Sell J. Physiological limitations and potential for improvement in gastrointestinal tract function of poultry. The Journal of Applied Poultry Research 1996; 5: 96-101

148. Serway RA, Jewett JW. Physics for scientists and engineers. 6 ed. Belmont: Brooks/Cole-Thomson Learning; 2004

149. Shamoto K, Yamauchi, K. Recovery responses of chick intestinal villus morphology to different refeeding procedures. Poultry Science 2000; 79: 718-723

150. Shan G. Biomechanical evaluation of bike power saver. Applied Ergonomics 2008; 39: 37-45

151. Silva JRL, Rabello CBV, Dutra Jr WM, Ludke MCMM, Barroso JF, Freitas CRG.; Campelo Filho EVB, Aquino LM. Efeito da forma física e do programa alimentar na fase préinicial sobre desempenho e características de carcaça de frangos de corte. Acta Scientiarum. Animal Sciences 2004; 26: 543-551

152. Skinner JT, Waldroup AL, Waldroup PW. Effects of dietary nutrient density on performance and carcass quality of broilers 42 to 49 days of age. The Journal of Applied Poultry Research 1992; 1: 367-372

153. Skinner-Noble DO, McKinney LJ, Teeter RG. Predicting effective caloric value of nonnutritive factors: III. Feed form affects broiler performance by modifying behavior patterns. Poultry Science 2005, 84: 403-411

154. Slater PJB. The temporal pattern of feeding in the Zebra finch. Animal Behaviour 1974; 22: 506-515

155. Smith M, Yanega GM, Ruina A. Elastic instability model of rapid beak closure in hummingbirds. Journal of Theoretical Biology 2011; 282: 41-51

156. Stansfield BW, Nicol AC, Paul JP, Kelly IG, Graichen F, Bergmann G. Direct comparison of calculated hip joint contact forces with

those measured using instru-mented implants. An evaluation of a three-dimensional mathematical model of the lower limb. Journal of Biomechanics 2003; 36: 929-936

157. Steeve RW. Babbling and chewing: jaw kinematics from 8 to 22 months. Journal of Phonetics 2010; 38: 445-458

158. Stefen C, Ibe P, Fischer MS. Biplanar X-ray motion analysis of the lower jaw movement during incisor interaction and mastication in the beaver (Castor fiber L. 1758). Mammalian Biology 2011; 76: 534-539

159. Svihus B. The gizzard: function, influence of diet structure and effects on nutrient availability. World›s Poultry Science Journal 2011; 67: 207-223

160. Svihus B, Klovstad KH, Perez V, Zimonja O, Sahlstrom S, Schuller RB. Physical and nutritional effects of pelleting of broiler chicken diets made from wheat ground to different coarsenesses by the use of roller mill and hammer mill. Animal Feed Science and Technology 2004; 117: 281-293

161. Thomas M, Van Der Poel AFB. Physical quality of pelleted animal feed. 1. Criteria for pellet quality. Animal Feed Science and Technology 1996; 61: 89-112

162. Thomas M, Van Vliet T, Van Der Poel AFB. Physical quality of pelleted animal feed 3. Contribution of feedstuff components. Animal Feed Science Technology 1998; 70: 59-78

163. Thomas M, Van Zuilichem DJ, Van Der Poel AFB. Physical quality of pelleted animal feed. 2. contribution of processes and its conditions. Animal Feed Science Technology 1997, 64: 173-192

164. Tran, HQ, Mehta RS, Wainwright PC. Effects oframspeedonpr eycapturekinematicsofjuvenileIndo-Pacific tarpon, Megalops cyprinoides. Zoology 2010; 113: 75-84

165. Truong TV, Le TQ, Byun D, Park HC, Kim M. Flexible wing kinematics of a free-flying beetle (*Rhinoceros Beetle Trypoxylus Dichotomus*). Journal of Bionic Engineering 2012; 9: 177-184

166. USDA, 2012. Livestock and Poultry: World Markets and Trade, s.l.: USDA.

167. Van Den Heuvel WF. Kinetics of the skull in chicken (*Gallus Gallus Domesticus*). Netherlands Journal of Zoology 1992; 42(4): 561-582

168. Van Der Heuvel WF, Berkhoudt H. Pecking in the chicken (*Gallus Gallus Domesticus*): Motion analysis and stereotypy. Netherlands Journal of Zoology 1998; 48(3): 273-303

169. Vargas GD, Brum PAR, Fialho FB, Rutz F, Bordin R.. Efeito da forma física da ração sobre desempenho de frangos de corte machos. Revista Brasileira de Agrociência 2001; 7(1): 42-45

170. Vestergaard K, Hogan JA, Kruijt JP. The development of a behaviour system: Dustbathing in the Burmese red junglefowl I. The influence of the rearing environment on the organization of dustbathing. Behaviour 1990; 112(1): 35-52

171. Vieira SL, Pophal S. Post-hatching nutrition in broiler. Brazilian Journal of Poultry Science 2000; 02(03): 189-199

172. Wainwright PC. Ecomorphology: experimental functional anatomy for ecological problems. American Zoologist 1991; 31: 680-693

173. Westneat MW, Long Jr JH, Hoese W, Nowicki S. Kinematics of birdsong: functional correlation of cranial movements and acoustic features in sparrows. The Journal of Experimental Biology 1993; 182: 147-171

174. Wolter BF, Ellis M, Curtis SE, Parr ENI, Webel DM.. Feeder location did not affect performance of weeling pigs in large groups. Journal of Animal Science 2009; 78: 2784-2789

175. Wroe S, Huber DR, Lowry M, Mchenry C, Moreno K, Clausen P, Ferrara TL, Cunningham E, Dean MN, Summers AP. Three-dimensional computer analysis of white shark jaw mechanics, how hard can a great white bite?. Journal of Zoology 2008; 276: 336-342

176. Wu G, Zeng L, Ji L. Measuring the Wing Kinematics of a Moth (Helicoverpa Armigera) by a Two-Dimensional Fringe Projection Method. Journal of Bionic Engineering 2008; 5: 138-142

177. Yoganandan N, Pintar FA, Cusick JF. Biomechanical analyses of whiplash injuries using an experimental model. Accident Analysis and Prevention 2002; 34: 663-671

178. Yo T, Vilariño M, Faure JM, Picard M. Feed pecking in young chickens: new techniques of evaluation. Physiology & Behavior 1997; 61: 803-810

179. Zang JJ, Piao XS, Huang DS, Wamg JJ, Ma X, Ma YX. Effects of Feed Particle Size and Feed Form on Growth Performance, Nutrient Metabolizability and Intestinal Morphology in Broiler Chickens. Asian-Australasian Journal of Animal Sciences 2009; 22(1): 107-112

180. Zeigler H, Levitt P, Levine R. Eating in the pigeon (*Columba livia*): movement patterns, stereotypy, and stimulus control. Journal of Comparative Physiology and Psychology 1980; 94: 783-794

181. Zweers GA. Pecking of the Pigeon (*Columba livia* L.). Behaviour 1982; 81(2/4): 173-230

Aggressive Behavior in the Genus Gallus Sp1

Queiroz SAI and Cromberg VUII

IDepartamento de Zootecnia, Faculdade de Ciências Agrárias e Veterinárias de Jaboticabal, Universidade Estadual Paulista (UNESP)

IIETCO – Grupo de Estudos e Pesquisa em Etologia e Ecologia - UNESP/FCAV

ABSTRACT

The intensification of the production system in the poultry industry and the vertical integration of the poultry agribusiness have brought profound changes in the physical and social environment of domestic fowls in comparison to their ancestors and have modified the expression of aggression and submission. The present review has covered the studies focusing on the different aspects linked to aggressiveness in the genus *Gallus*. The evaluated studies have shown that aggressiveness and subordination are complex behavioral expressions that involve genetic differences between breeds, strains and individuals, and differences in the cerebral development during growth, in the hormonal metabolism, in the rearing conditions of individuals, including feed restriction, density, housing type (litter or cage), influence of the opposite sex during the growth period, existence of hostile stimuli (pain and frustration), ability to recognize individuals and social learning. The utilization of fighting birds as experimental material in the study of mechanisms that have influence on the manifestation of aggressiveness in the genus *Gallus* might comparatively help to elucidate important biological aspects of such behavior.

INTRODUCTION

Domestic fowls are social birds that live naturally in groups constituted by a rooster with many hens in a determined territory. Group structure is hierachical with male dominance over all hens and, among these, social hierarchy is determined by pecking order and perch location. Such hierarchical system is rigorously maintained by aggressive behaviors, but once the hierarchy has been established, the aggressiveness decreases and is substituted for demonstrations of dominance (threatening) and submission (subordination). Roosters and hens form separate social hierarchies. The dominance of males over females is seldom contested, so that aggressive male behavior towards hens is hardly seen under natural conditions (Milman and Duncan, 2000). On the other hand, aggression between males during the reproductive season is fairly common due to the increase in testosterone levels (Ros *et al.*, 2002) and competition for mating opportunities.

The physical aspects of rearing environment and social experience might have important roles in the development and expression of agonistic behaviors by captive animals. There is an increase in the threshold of defensive behavior responses with domestication, resulting in captive animals that show lower submission response or social inhibition in comparison to wild animals (Lorenz, 1950). Selection against defensive behavior might represent an adaptation in order to minimize stress of inhabiting an environment where it is not possible to escape from aggression.

Evidences indicate that one of the most important effects of domestication on animal behavior is the reduction of emotional

Selection modifies the sensitivity to androgens. Ortman and Craig (1968) selected for high or low social dominance within groups of White Leghorn and Red Rhode Island chickens during five generations. Comparison was performed using four-month-old males from many strains that had been castrated before two weeks of age. The males were sub-divided into groups and androgens were administered at different doses. Assessments were performed by means of competition between two capons of each strain selected for high and low social dominance that were given the same androgen level. Two conclusions were taken: **(1)** androgen administration increased agonistic behavior in castrated males, thus without the testicular source of male hormone; and (2)males from the strains selected for greater aggressiveness showed more probability of winning competitions against the males of less aggressive strains. The changes produced by genetic selection were caused mainly through the changes in the physiological responses, and not due to changes in the amount of androgen that was produced.

The problems caused by aggressiveness in the poultry industry are more associated to laying hens, whereas problems with passive and docile behaviors have been generally reported in broiler breeders (Milman and Duncan, 2000). Nevertheless, such differences are probably more related to differences in maturity than genetics, since broiler breeders are marketed at approximately 42 days, i.e., they are still young. At this age, meat-type roosters showed lower aggressiveness than egg-type roosters (Mench, 1988), but it is possible that aggressiveness in broiler chickens does not increase before the sexual maturity.

According to Milman and Duncan (2000), broiler breeder producers have been facing problems of increasing aggressiveness in the rearing phase and, more recently, also in the reproductive phase. Males have shown extreme aggressiveness during mating, forcing copulation and causing serious injuries to females or even death. Some males chase females and trap them in the corners of the poultry houses. Frightened females scape, hide in the nests and avoid the males. Therefore, they stay in less attractive areas in terms of food and water, which diminishes the flock fertility and consequently causes economical losses. In order to explain such observations, the differences in the sexual behavior have been evaluated in terms of rooster aggressiveness in three different strains: a broiler strain, a layer strain and a fighting rooster strain (Milman and Duncan, 2000). There were different levels of bird management during the experiment, i.e., fighting roosters were less docile and showed tendency to fight with other males. Nevertheless, the aggressive behavior of the male against females was reported only in the broiler strain. In the other two strains, the males elaborated all the expected reproductive behaviors and copulation, without threatening or harming the females. In the broiler strain, there were abnormalities on the sequence of reproductive behaviors of males, i.e., some of them have ignored the courting behavior, beginning directly with climbing and copulation. Females, in turn, were not aware of male intentions and would not be prepared for mating nor demonstrate the expected behaviors of crouching and exposure of the cloaca. The alterations of male reproductive behavior in this strain might have been an indirect response to the intense selection for higher growth rates and development of breast and thigh muscles.

Stress and Aversion Stimuli

Aggression is highly related to responses to pain in almost all species (Craig, 1981) and, it generally leads to aggression and fights if induced in the presence of partners (Ulrich, 1966).

Frustration, such as that caused by feed restriction in hens, may also trigger aggressiveness. Frustration was inflicted to hungry birds by covering the feed tray with clear plastic, so that the food was seen but could not be touched (Duncan and Wood-Gush, 1971). In every tested situation, the frustrated birds that were socially dominant showed increased aggression againt submissive birds, and aggression frequency increased considerably when feed restriction lasted the whole day in comparison to shorter periods of restriction.

Males generally exibit "passive dominance" over females and are rarely caught pecking or threatening females under everyday circumstances. Nevertheless, hungry and frustrated broilers were excessively aggressive when paired with females under the conditions of the above-mentioned study (Duncan and Wood-Gush, 1971). Females were pecked 806 times or threatened by the hungry and frustrated birds in 8 hours, in comparison to 18 peckings when kept with non-frustrated males.

King (1965) has shown that extremely frustrating situations might generate questionable results, particularly when social tension levels are already high, such as in the case of young broilers. The author has determined the social hierarchy by the pecking order in three groups of young broilers and compared the frequency of aggression and the estability of dominance relationships after restricting the birds for 24 hours and then showing the food for one hour in three different manners. Uniform spread of the food on litter resulted in very low aggression frequency, since all individuals were busy eating during that period. Food provided in a round feeder that permitted the access of all individuals, but forming groups, increased aggressions 36 times in comparison to the previous situation, and the incidence of attack by subordinates or threatening dominant birds was 5%. Finally, individual access to food caused an outburst of aggressions and violations of the pecking order that happened at a frequency of approximately 50%. The established dominance relationships returned to normal values in the absence of extremely frustrating situations.

Craig (1981) has suggested many possibilities for aggression when the animals are frustrated and, probably, more than one would be happening at the same time. Frustration occurs together with excitation and consequent movimentation. The consequent changes in the activity patterns of individuals generally leads to more frequent interactions than usual. In case that there is a single and scarce resource, the personal space of dominant individuals might be invaded by subordinates and generate an aggressive response, or the situation might be so hostile that would trigger the agonistic behavior, such as in the case of pain.

When aggression is rewarded, fighting might become a means of satisfying a necessity. Once aggression begins, it might continue for longer than the time needed to obtain a reward, although the extinction of fight response might probably happen (Craig, 1981).

Many aspects of rooster fight might be explained by rewards paired with different responses, since growing to training, until fighting. Reward might be positive, represented by stimuli with positive connotations, such as a resource indispensable to survival or with hedonistic values, such as water, feed and company, or yet negative, associated to the absence of electrical shocks or other stimuli with aversive connotation.

Individual Space and Territorial Behavior

Aggression or aggression threatening are commonly used to exclude individuals from the personal space of an animal. McBride *et al.* (1963) have shown that the personal space do not extends equally in every direction, but it was bigger in the front of the chickens and most of the movements were intended to avoid the personal spaces of dominant hens. Attacks or threatening behaviors were ended frequently by the subordinate escaping to the limit of the personal space of a dominant animal, or by submission postures or out-of-order behaviors. These are behaviors out of the context of the situation, such as asking for food or displaying sexual facilitation behavior.

Many species, among which the domestic fowl, allows the subordinates to stay in the group as long as they exibit submission behaviors, but subordinates might sometimes be hurt or expelled (Craig, 1981; Price, 2002). The inability to escape or the absence of a submission behavior might cause the death in fighting roosters. The personal space may be minimized in non-competitive situations, but increases when resources are limited.

There are two typical responses of individuals to others of the same species, which is dependent on if they belong or not to the group (Craig, 1981). Although personal space is also present in groups, the existence of social distance indicates cohesion. Adult members in the group normally react with aggressive behaviors against intruders. As a consequence of taming, natural social groupings are rarely permitted. Few males are kept intact, the young are separated from their dams very early and, periodically, the individuals are re-organized in new groups. Nevertheless, the knowledge of how the social groups behave and organize themselves in the space might provide valuable information to understand behavior problems that are seen under artificial rearing conditions.

McBride *et al.* (1969) have described the territorial behavior of the red jungle fowl in the Southern Asian forest, which may be seen before and after the reproductive season. Two other manifestations of this behavior have been observed. In one situation, hens incubating eggs and those with young chicks became solitary and occupied living areas (home range). Such areas were overlapped for the different birds, which resulted in dominance relationships between pairs of molting hens or with progeny. Thus, although the two hens could live in the same area, they would tipically avoid each other. On the other situation, dominant males would stay in fixed areas out of the reproductive season, but did not have exclusive control on such areas. Subordinate males moved between the limits of the groups and stayed at the external areas of the groups to which they were joined.

The dominant rooster has a fundamental role on the determination of the group movements and vigilance for intruders. McBride et al. (1969) observed that when the group moved to another territory, the male was the one that gathered the females before moving. The females interacted with the rooster while in movement, and the male had control over the space when they crossed areas without vegetation. In the case of threats, the male alerted the females and would walk parallel to the predator or potential predator, whereas the females would hide quietly. When the group was threatened, the male frightened the females by running towards them with opened wings. The male would be most of the time on guard and show an alert position while females ate, raising the tail and lowering the wings. On the other hand, the females were closer to the male while he relaxed to eat. The male was generally much more cautious than the females. It protected the females from other males, and threatened the intruder. Molting females and hens with chicks controlled the movements of the group, called attention to the food, were vigilant for intruders and defender in the presence of a potential danger.

Effect of Isolation and Overcrowding on Aggressiveness

Under some circumstances, the animals are attracted to each other and maintain a close social interaction as a function of the space (e.g., the dam and its progeny, males and females during the reproductive season, and search for shelter close to the body of others); however, it is also observed that there must be a minimum space between individuals.

Birds reared in isolation show precocious and more intense aggressiveness than those reared in groups (Guhl, 1953, 1958; Gulh et al., 1960). Increased density, for example, housing of 100 individuals in an area previously occupied by 25, multiplies the group size by four and decreases the area per animal to a fourth of the previously available area. Decreasing the area per bird in half might affect groups of 4 or 400 individuals very differently.

Many studies with hens have suggested an interesting relationship between the area per bir and aggression frequency. The decrease in the available space increased the frequency of aggressions and there is then a marked decrease as a function of greater agglomeration (Al-Rawi and Craig, 1975).

Hughes and Wood-Gush (1977) have also found a marked decrease in aggression under conditions of high density of caged birds. The observations indicated that normal displaying of threatening require a minimal area per bird that is not possible in the majority of the cages. Studies by Bhagwat and Craig (1979) have also clearly evidenced the reduction in the incidence of cage threatening. There was a reduction in aggressive peckings directed towards the head in ambients with high density. This might be explained by the fact that, under situations of extreme agglomeration, the pecking mechanism of a dominant bird is not activated or triggered by subordinate chickens, if these are already within the influence area of the dominant bird. Only the entrance into the personal space of an individual would cause such behavior (Hughes and Wood-Gush, 1977).

Results obtained by Ylander and Craig (1980) have demonstrated that the socially dominant bird (the third part or the third member) inhibits the aggressive interactions between pairs of subordinate birds. Males were particularly effective as inhibitors of aggressive behavior between dominant members of pairs of hens during feeding behaviors. There were only five escapes from pecking in 24 tests of ten minutes with hen pairs close to a male; when the male was one meter away, there were 21 and, when the male was temporarily removed from the area, there were 74 escapes. Escaping from threatening followed the same pattern, but the effects were less evident. Therefore, female aggressiveness was reduced in the presence of males in large or small groups (McBride et al., 1969; Craig and Bhagwat, 1974; Bsary and Lamprecht, 1994). It is not clear, however, whether this effect might be attributed to the dominance of males or because sub-group formation is facilitated (Odén et al., 2000).

Pairs of males kept in large cages with solid back walls from 12 to 20 weeks of age were easily classified into dominant and subordinate before 20 weeks, since the subordinate showed clear signs of physical abuse and submission posture, giving indirect evidence of frequent and severe aggression by the dominant member (Grosse and Craig, 1960). In the birds placed in cages that permitted 30% of this space per bird (Craig and Polley, 1977), the subordinate males delayed sexual maturity, but pairs of males in the second trial did not show any sign of physical abuse. Probably, aggression between individuals was not possible due to the reduced available space.

Effect of Group Size on Aggressiveness

Guhl (1953) has provided partial evidence that a group with 96 birds showed a complete pecking order. In a later study, groups with 100 to 400 birds were observed (Craig and Guhl, 1969). In the groups with 200 individuals, the hens tended to stay longer in some areas and were dominant in these places. Probably, the fixation to particular areas limits the necessity of recognizing a larger number of hens. Thus, the social difficulties associated to the gathering of many intruders in bigger groups might be prevented by the tendency of hens to fix in their own neighborhood.

Although groups of 100 or more birds might be socially organized, Banks (1956) has observed that the violations of the pecking order were more frequent when group size increased from 6 to 24 birds. A violation of the pecking order consists in a subordinate attacking its social superior. The relative frequency of such behavior has been associated to the group size, and more violations were seen in larger groups. The violations, however, were inefficient, since insubordination was immediately retaliated, and there has not been any reversion of the dominance degree. The author suggested that the reinforcement represented by the social position occupied in particular by hens was less important in larger groups, indicating that the limits for recognition were being reached. Evidences of other studies indicated that the pecking order is relatively stable in larger groups than those mentioned herein, nevertheless temporary confusion and lack of recognition might be responsible for higher aggression levels in larger groups.

Al-Rawi and Craig (1975) evaluated agonistic behavior and reported that the frequency of aggressive acts by hens housed in cage batteries increased with the increase in the group size (4 to 28 birds). Most of the described aggressive acts were pecking instead of threatening, and these have happened during feeding or when birds approached the feeder. In groups of 4, 8 and 14 birds, Al-Rawi et al. (1976) observed higher levels of aggression in the larger groups in the first eight weeks, but agonistic behavior was reduced in all groups, before being observed again at 26 weeks after grouping. Possibly, the hens that must live too close to each other in cages become so familiarized that there are not temporary failures of recognition, even in groups of 8 and 14 birds.

Estevez et al. (2003) have studied the ontogenesis of the aggressive behavior in chickens from 3 to18 weeks of age. It was suggested that the birds establish hierarchical dominance through aggressive interactions in small groups, but larger groups adopt a more tolerant social strategy and aggressiveness is reduced. Focal observations of birds within groups showed a decrease in the frequency of peckings and threatenings with the increase in the group size, although pecking and threatening frequency has increased with group size for some birds. Therefore, it becomes evident that hens adopt different social strategies with the increase in the size of groups, and it might be speculated that most birds might adopt a strategy of tolerance in large groups, whereas a minority might be tyrannical, addressing aggression indiscriminately towards other birds.

The system of social dominance that is constructed based on fighting and memory recognition of the individual positions in large groups is not sustained (D'Eath and Keeling, 2003). Besides, hens adapt and become less aggressive or restrict the movements towards defined territories. Some evidences have also indicated that the hens in large groups have not established the territory inside a separated area. Among others, Odén et al. (2000) identified the existence of sub-groups associated to territories when the hens had to live in large groups and attributed the difficulty of the observer in identifying an individual bird to the failure in recognizing such sub-groups. These overall findings corroborate results reported by Pagel and Dawkins (1997), who observed that hens in large groups are less aggressive and might change the social system to a system in which hierarchy is directly determined through the access to dominance and subordination signs instead of individual recognition in the small group.

Aggression Maintenance

Individuals placed together in an unfamiliar area usually inspect the new ambient and partners, and usually there is quietness. Then, one or two pairs begin to interact agonistically. Males might be involved in fightings, whereas the females interact less vigorously in the majority of the hen breeds. Breeds and strains might differ in the intensity and duration of competitions for a dominance position (Craig, 1981). Fighting roosters might fight to death, unless they are separated; nevertheless, in most groups the result is decided rapidly within pairs. If the number of grouped birds is relatively small, the dominance order of all possible pairs might be established within some hours, although agonistic interactions might continue for relatively longer periods.

It is suggested that aggression might happen between young animals because they start playing spontaneously and playing become more vigorous with aging, until pain is accidently inflicted by one of the birds. It is known that pain triggers a defensive behavior or reflexive fight. The conditioning would explain the tendency to establish immediately dominance relationships; the dominant individual is rewarded and the subordinate is rewarded by the submissive behavior, since it will no longer be attacked.

Lorenz (1950) suggested the existence of a "specific energy for aggression", which would spontaneously begin in the nervous system and would accumulate until a limit, after which it should be released, in an analogy to a water source that fills a vessel and, once a limit is reached, the water should be discharged. On the other hand, Scott (1971) has considered fear and anger as the primarily responsible emotions for social fighting, which would be triggered by external stimuli, but once they had been activated, they would be extended and would increase reactivity to external stimuli, particularly if the behavior in curse was blocked. In the absence of additional external excitement, internal stimulations associated with anger and fear would be extinguished.

Some Considerations

In the different *Gallus* species, natural selection has acted in favor of birds that were more efficient in acquiring and maintaining exclusive territories that provided them with abundant feeding, shelter and protection against predators. On the other hand, natural selection has also favored an intense internal competition among individuals of a population, mainly competition for mating opportunities. In this scenery of constant competition for territorial maintenance and group exclusiveness, dominant roosters with attributes that enable them to intense vigilance, prevention against invations (threatening displays) and combat habilities had more descendants and increased the presence of its genes in the population. Similar patterns of vigilance and defense are exhibited by females imbued with care to the progeny. Therefore, it is easy to understand the importance that the aggressive behavior and the fighting abilities have on this species under natural conditions.

The same peculiarities of the hierachical social system, together with the promiscuous reproductive behavior, the extensive feeding habits and the short life cycle of these animals, made domestication easy and enabled prompt world dispersion of these species in the genus *Gallus*. Afterwards, the birds have been submitted to different rearing systems and artificial selection of the best individuals in those specific environments, resulting in a wide range of breeds and, later, strains within breeds that have been developed mainly for human feeding. Selection of production attributes has certainly favored the permanence of animals that were more docile and with greater social tolerance.

Poultry rearing has suffered profound changes after the mid 1900s, and has become an activity in which the main decisions concerning selection and reproduction were controlled by few commercial companies. It was then called poultry industry, in which there was intensification of the production systems, smaller area per bird, higher number of birds housed in the same area, preventive administration of drugs and search for management that minimized the problems that resulted from the new rearing conditions. The majority of the reviewed literature from the second half of the 20th century deals with finding out and proposing solutions to the behavioral problems caused by the intensification of the rearing system. The virtual absence of genetic studies on behavior in the same period reflects in part the fact that some companies were not interested in revealing the advances and problems faced by them neither their search for solutions to the problems that have emerged, since there was no interest in modifying the intensive and vertical production system. However, the few genetics studies of aggressiveness behaviors in domestic fowls evidenced that neither domestication nor the intense artificial selection were able to change the social behavior of the birds, and that the intrinsic patterns of aggressiveness of this species are still seen.

It is also noticeable the lack of scientific studies with fighting breeds of birds. It is expected that the individuals from these breeds show the same biological mechanisms and aggressive behavior patterns that their conspecifics. On the other hand, these are possibly exacerbated in fighting birds, since they have not undergone the attenuating effects of artificial selection that is practiced in the commercial meat-type or egg-type strains of birds. Besides, the rearing environment of fighting birds is more close to natural habitats, i.e., both incubation and growth happen in the presence of the dam and the other birds of the group. Therefore, the study of aggressiveness expression in the birds that are closer to the original wild genotype would enable a broader comprehension of the biological mechanisms involved in such behaviors, as well as the importance and usefulness of these birds to the current animal populations.

REFERENCES

1.   Adret-Hausberger M, Cumming RB. Social attraction to older birds by domestic chickens. Bird Behaviour 1987; 7:44-46.

2.   Al-Rawi B, Craig JV. Agonistic behavior of caged chickens related to group size and area per bird. Applied Animal Ethology 1975; 2:69-80.

3.   Al-Rawi B, Adams AW, Craig JV. Agonistic behavior and egg production of caged layers: genetic strain and group-size effects. Poultry Science 1976; 55:796-807.

4.   Andrew RJ. Effects of testosterone on the behaviour of the domestic chick. I. Effects present in males and not in females.

Animal Behavior 1975a; 23:139-155.

5.   Andrew RJ. Effects of testosterone on the behaviour of the domestic chick. II. Effects present in both sexes. Animal Behavior 1975b; 23: 156-158.

6.   Astiningsih K, Rogers LJ. Sensitivity to testosterone varies with strain, sex, and site of action in chickens. Physiology and Behavior 1996; 59:1085-1092.

7.   Banks EM. Social organization in red jungle fowl hens (*Gallus gallus subsp*). Ecology 1956; 37(2):239-248.

8.   Bardo P. Central nervous mechanism for expression of anger in animals. In: Reymart ML editor. Feelings and emotions. New York: McGraw-Hill; 1950.

9.   Bateson PPG. Changes in chicks' response to novel moving objects over the sensitive period of imprinting. Animal Behavior 1964; 12: 479-489.

10.   Bateson PPG. How do sensitive periods arise and what are they for? Animal Behavior 1979; 27:470-486.

11.   Bateson PPG. Is imprinting such a special case? Philosophical Transactions of the Royal Society of London 1990; 329:125-131.

12.   Bhagwat AL, Craig JV. Effects of male presence on agonistic behavior and productivity of White Leghorn hens. Applied Animal Ethololology 1979; 5:267-282.

13.   Bshary R, Lamprecht J. Reduction of aggression among domestic hens (*G. domesticus*) in the presence of a dominant third party. Behaviour 1994; 128:311-324.

14.   Bullock SP, Rogers LJ. Sex differences in the effects of testosterone and its metabolites on brain asymmetry for the control of copulation in young chicks. In: 16th Proceedings of the Australian Physiology and Pharmacological Society; 1985; Sydney, Australia. p.235.

15.   Craig JV, Guhl AM. Territorial behavior and social inter-actions of pullets kept in large flocks. Poultry Science 1969; 48:1622-1628.

16.   Craig JV, Bhagwat AL. Agonistic and mating behavior of adult chickens modified by social and physical environments. Applied Animal Ethology 1974; 1:57-65.

17.   Craig JV, Polley CR. Crowding cockerels in cages: effects on weight gain, mortality and subsequent fertility. Poultry Science

1977; 56:117-120.

18. Craig JV. Domestic animal behavior. New Jersey: Pretince-Hall; 1981.

19. D'Eath RBD, Keeling LJ. Social discrimination by laying hens in large groups: from peck orders to social tolerance. Applied Animal Behavior Sciences 2003; 84:197-212.

20. Dimond SJ. Effects of photic simulation before hatching on the development of fear in chicks. Journal of Comparative and Physiological Psychology 1968; 65:320-324.

21. Duncan IJH, Wood-Gush DGM. Frustration and aggression in the domestic fowl. Animal Behavior 1971; 19:500-504.

22. Estevez I, Keeling LJ, Newberry RC. Decreasing aggression with increasing group size in young domestic fowl. Applied Animal Behavior Sciences 2003; 84:213-218.

23. Gottlieb G. Imprinting in relation to parental and species identification by avian neonates. Journal of Comparative Physiology and Psychology 1965; 59: 345-356.

24. Grosse AE, Craig JV. Sexual maturity of males representing twelve strains of six breeds of chickens. Poultry Science 1960; 39:164-172.

25. Guhl AM. Social behavior of domestic fowl. Kansas: Kansas Agricultural Experimental Station; 1953. (Bulletin, 73).

26. Guhl AM. The development of social organization in the domestic chick. Animal Behaviour 1958; 6:92-111.

27. Guhl AM, Craig JV, Mueller CD. Selective breeding for aggressiveness in chickens. Poultry Science 1960; 39:970-980.

28. Hess EH. Imprinting. Science 1959; 130:133-141.

29. Hughes BO, Wood-Gush DGM. Agonistic behaviour in domestic hens: the influence of housing method and group size. Animal Behaviour 1977; 25:1056-1062.

30. Hutchison JB, Steimer TJ, Hutchison RE. Formation of behaviorally active oestrogen in the dove brain: induction of preoptic aromatase by intracranial testosterone. Neuroendocrinology 1986; 43:416-427.

31. Johnson MH, Bolhuis JJ, Horn G. Predispositions and learning: Behavioral dissociation in the chick. Animal Behaviour 1992; 44: 943-948.

32. Jones RB. Sex and strain differences in the open-field responses of the domestic chick. Applied Animal Ethology 1977; 3:255-261.

33. Jones RB, Waddington D. Modification of fear in domestic chicks, *G. gallus domesticus*, via regular handling and early environmental enrichment. Animal Behaviour 1992; 43:1021-1033.

34. Jones RB, Waddington D. Attenuation of the domestic chick's fear of human beings via regular handling: in search of a sensitive period. Applied Animal Behaviour Science 1993; 36:185-195.

35. Kilham P, Klopfer PH, Oelke H. Species identification and color preferences in chicks. Animal Behaviour 1968; 16:238-244.

36. King MG. Disruptions in the pecking order of cockerels concomitant with degrees of accessibility to feed. Animal Behaviour 1965; 13:504-506.

37. Komai T, Craig JV, Wearden S. Heritability and repeatability of social aggressiveness in the domestic chicken. Poultry Science 1959; 38(2): 356-359.

38. Kruijt JP. Ontogeny of social behaviour in Burmese Red Jungle Fowl (*Gallus gallus spadiceus*). Behaviour 1964; suppl 12:1-201.

39. Leonard ML, Zanette L, Fairfull RW. Early exposure to females affects interactions between male White Leghorn chickens. Applied Animal Behaviour Science 1993; 36:29-38.

40. Loiselet J. Behaviour and feather pecking are priority areas for selection. World Poultry 2004; 7(20):22-23.

41. Lorenz KZ. The comparative method in studying innate behaviour patterns. Symposium of Society of Experimental Biology 1950; 4: 221-269.

42. McBride G, James JW, Shoffner RN. Social forces determining spacing and head orientation in a flock of domestic hens. Nature 1963; 197:1272-1273.

43. McBride G, Parer IP, Foenander F. The social organisation of behaviour of the feral domestic fowl. Animal Behaviour Monograph 1969; 2: 127-181.

44. McDaniel GR, Craig JV. Behavior traits, semen measurements and fertility of White Leghorn males. Poutry Science 1959; 38(5):1005-1014.

45. Mench JA. The development of aggressive behaviour in male broiler chicks: a comparison with laying-type males and the effects of feed-restriction. Applied Animal Behaviour Sciences 1988; 21:233-242.

46. Milman ST, Duncan IH. Strain differences in aggressiveness of male domestic fowl in response to a male model. Applied Animal Behaviour Sciences 2000; 66:217-233.

47. Newman S. Quantitative and molecular genetic effects on animal well-being adaptive mechanisms. Journal of Animal Science 1994; 72:1641-53.

48. Oden K, Vestergaard KS, Algers B. Space use and agonistic behaviour in relation to sex composition in large flocks of laying hens. Applied Animal Behaviour Sciences 2000; 67:307-320.

49. Ortman LL, Craig JV. Social dominance in chickens modified by genetic selection-physiological mechanisms. Animal Behaviour 1968; 16:33-37.

50. Pagel M, Dawkins MS. Peck orders and group size in laying hens: future contracts for non aggression. Behavioural Processes 1997; 40:13-25.

51. Phillips RE, Youngren OM. Unilateral kainic acid lesions reveal dominance of right archis triatum in avian fear behavior. Brain Research 1986; 377:216-220.

52. Phillips RE, Siegel PB. Development of fear in chicks of two closely related genetic lines. Animal Behaviour 1966; 14:84-88.

53. Price EO. Animal domestication and behavior. Cambridge: Cab International; 2002.

54. Rajecki DW. Formation of leap orders in pairs of male domestic chickens. Aggressive Behavior 1988; 14:425-436.

55. Regolin L, Vallortigara G, Zanforlin M. Perceptual and motivational aspects of detour behaviour in young chicks. Animal Behaviour 1994; 47:123-131.

56. Rogers LJ, Workman L. Light exposure during incubation affects competitive behaviour in domestic chicks. Applied Animal Behaviour Science 1989; 23:187-198.

57. Rogers LJ, Astiningsih K. Social hierarchies in very young chicks. British Poultry Science 1991; 32:47-56.

58. Rogers LJ. The development of brain and behaviour in the chicken. Cambridge: CAB International;1995.

59. Rogers LJ, Zappia JV, Bullock SP. Testosterone and eye-brain asymmetry for copulation in chickens. Experientia 1985; 41:1447-1449.

60. Ros AFH, Dieleman SJ, Groothuis TGG. Social stimuli, testosterone, and aggression in gull chicks: support for the challenge hypothesis. Hormones and Behaviour 2002; 41:334-342.

61. Rushen J. Frequencies of agonistic behaviours as measures of aggression in chickens: a factor analysis. Applied Animal Behaviour Science 1984; 12:167-176.

62. Salzen EA. Imprinting and fear. Zoological Society of London Symposium 1962; 8:199-217.

63. Scott JP. Theoretical issues concerning the origin and causes of fighting. In: Eleftheriou BE, Scott JP, editors. The Physiology of aggression and defeat. New York: Plenum Press; 1971.

64. Siegel PB. Evidence of a genetic basis for aggressiveness and sex drive in the White Plymouyh Rock Cock. Poultry Science 1959; 38(1):115-118.

65. Sluckin W. Imprinting and learning. London: Methuen; 1966.

66. Tanabe Y, Nakamura T, Fujioka K, Doi O. Production and secretion of sex steroid hormones by testes, the ovary and the adrenal glands of embryonic and young chickens (G. domesticus). General and Comparative Endocrinology 1979; 39:26-33.

67. Tindell D, Craig JV. Genetic variation in social aggressiveness and competition effects between sire families in small flocks of chicken. Poultry Science 1959; 39:1318-20.

68. Tuculescu RA, Griswold JG. Prehatching interactions in domestic chicks. Animal Behavior 1983; 31:1-10.

69. Ulrich R. Pain as a cause of aggression. American Zoologist 1966; 6: 643-662.

70. Vallortigara G, Andrew RJ. Lateralization of response by chicks to change in a model partner. Animal Behavior 1991; 41:187-194.

71. Vallortigara G, Andrew RJ. Olfactory lateralization in the chick. Neuropsychologia 1994; 32:417-423.

72. Vallortigara G. Affiliation and aggression as related to gender in domestic chicks (*Gallus gallus*). Journal of Comparative Psychology 1992; 106:53-57.

73. Vidal JM. The relations between filial and sexual imprinting in the domestic fowl: effects of age and social experience. Animal Behavior 1980; 28:880-891.

74. Workman L, Andrew RJ. Simultaneous changes in behaviour and lateralization during the development of male and female domestic chicks. Animal Behavior 1989; 88:596-605.

75. Ylander DM, Craig JV. Inhibition of agonistic acts between domestic hens by a dominant third party. Applied Animal Ethology 1980; 6: 63-69.

76. Young CE, Rogers LJ. Effects of steroidal hormones on sexual, attack, and search behavior in the isolated male chick. Hormones and Behavior 1978; 10:107-117.

77. Zajonc RB, Wilson WR, Rajecki DW. Affiliation and social discrimination produced by brief exposure in day old domestic chicks. Animal Behavior 1975; 23:131-138.

78. Zayan R. An analysis of dominance and subordination experiences in sequences of paired encounters between hens. In: Zayan R, Duncan IJH, editors. Cognitive aspects of social behaviour in the domestic fowl. Amsterdan: Elsevier; 1987.

Citations

CHAPTER 1

Atsushi Hirao, "The Possible Role of the Uropygial Gland on Mate Choice in Domestic Chicken," International Journal of Zoology, vol. 2011, Article ID 860801, 5 pages, 2011. doi:10.1155/2011/860801.

CHAPTER 2

Daniel Nätt, Carl-Johan Rubin, Dominic Wright, Martin Johnsson, Johan Beltéky, Leif Andersson, and Per Jensen, Heritable Genome-Wide Variation of Gene Expression and Promoter Methylation between Wild and Domesticated Chickens, doi:10.1186/1471-2164-13-59.

CHAPTER 3

A. C. Davies, A. N. Radford, I. C. Pettersson, F. P. Yang, and C. J. Nicol, Elevated arousal at time of decision-making is not the arbiter of risk avoidance in chickens, doi: 10.1038/srep08200.

CHAPTER 4

Bart Buitenhuis, Jakob Hedegaard, Luc Janss, and Peter Sørensen, Differentially Expressed Genes for Aggressive Pecking Behaviour in Laying Hens, doi:10.1186/1471-2164-10-544.

CHAPTER 5

Cinzia Chiandetti, Jessica Galliussi, Richard J. Andrew, and Giorgio Vallortigara, Early-Light Embryonic Stimulation Suggests a Second Route, Via Gene Activation, to Cerebral Lateralization in Vertebrates, doi:10.1038/srep02701.

CHAPTER 6

Haiping Xu, Xu Shen, Min Zhou1, Meixia Fang, Hua Zeng, Qinghua Nie, and Xiquan Zhang, The genetic effects of the dopamine D1 receptor gene on chicken egg production and broodiness traits, doi:10.1186/1471-2156-11-17.

CHAPTER 7

Neves DP;Banhazi TM;Nääs IA, Feeding Behaviour of Broiler Chickens: a Review on the Biomechanical Characteristics, doi. org/10.1590/1516-635x16021-16

CHAPTER 8

Queiroz SA and Cromberg VU, Aggressive Behavior in the Genus Gallus Sp1, http://dx.doi.org/10.1590/S1516-635X2006000100001.

Index